AF674805

PHARMACEUTICAL APPLICATIONS OF *Biocatalysis*

PHARMACEUTICAL APPLICATIONS OF *Biocatalysis*

Editor
R.C. Tripathi

GENE TECH

2006

Gene-Tech Books
New Delhi - 110 002

2006, © Publisher

Information contained in this work has been published by Gene-Tech Books and has been obtained by its author(s)/editor(s) from sources believed to be reliable and are correct to the best of their knowledge. However, the publisher and its author(s) make no representation of warranties with respect of accuracy or completeness of the contents of this book, and shall in no event be liable for any errors, omissions or damages arising out of use of this information and specifically disclaim any implied warranties or merchantability or fitness for any particular purpose.

All rights reserved. Including the right to translate or to reproduce this book or parts thereof except for brief quotations in critical reviews.

ISBN 81-89729-04-7

Published by : **GENE-TECH BOOKS**
4762-63/23, Ansari Road, Darya Ganj,
NEW DELHI - 110 002
Phone: 41562849
e-mail: genetechbooks@yahoo.co.in

Printed at : **Tarun Offset Printers**
Delhi

PRINTED IN INDIA

Preface

Biocatalysis is the use of an enzymatically-catalyzed reaction to convert a single starting compound to a single product, without consumption of the substrate for production of biomass, energy, or waste. Biocatalyst technology is an important tool for chemical synthesis. It has a large impact in the pharmaceutical world. It offers numerous advantages for manufacturing pharmaceuticals with unique and highly desirable characteristics and offers new production opportunities for a wide range of drugs.

The present book describes the pharmaceutical applications of biocatalysis in detail. It nicely brings together the essentials of biocatalysis including reactions, the products and processes that utilize the methodology, and techniques for improving biocatalysts. This book will be a useful reference tool for anyone interested in the fields of biocatalysis and pharmaceuticals, whether they be graduate students or more experienced practitioners.

Editor

Contents

1

Pharmaceuticals

The impact of biotechnology on pharmaceuticals is mainly based on biotechnical methods that can be applied in research, development and manufacturing. The market share of biopharmaceuticals started from a low level in the early 1990's, but is rapidly increasing in the international medical, drug market. Biopharmaceuticals can be expected to constitute about 10 % of the pharmaceutical market at the end of the decade. Most biopharmaceuticals are based on large and heavy molecules, often proteins or glycoproteins, while many traditional synthetic drugs are products of low molecular weight. The methods used in the identification, development, production, and delivery of biopharmaceuticals often differ dramatically from those in traditional pharmacology.

Drug delivery is a typical problem of biopharmaceuticals. It is not easy for large molecules to penetrate various chemical barriers in the body. That is why efforts to develop biopharmaceuticals need to be supported at an early stage with studies on means of efficient drug delivery. Because of problems of drug delivery, many biopharmaceuticals may eventually be replaced by synthetic small-molecule compounds. In many cases, though, the replacements will take advantage of the particular shape of the pharmacologically active region of the original biopharmaceutical molecule.

Biopharmaceuticals

The first scientific drug discovery efforts concentrated on evaluating natural compounds that might be used in pharmaceuticals; the emphasis was on extracting beneficial biooriginated ingredients from plants and other sources. During the first half of the 20th century, systematic pharmacological evaluations of both synthetic and natural compounds were carried out. After the usefulness of a candidate compound was ascertained, a synthetic process was developed for its production. Advances in antibody biology and gene technology placed the drug discovery process on a more rational footing in the 1980's. It became possible to design drugs for specific molecular receptor sites and protein abnormalities, including deficiencies in enzymes.

The development of computerised methods has further increased the opportunities for rational drug design. The power of the rational approach is confirmed by the increasingly accurate biomolecular explanations discovered for the efficacy of existing, experimentally developed drugs. For instance, it has been discovered that the reason why aspirin helps to reduce inflammation is that it protects a protein that inhibits the onset of the inflammatory process. Such explanations help to develop further and ascertain rational drug design principles. It has been estimated that there are already more than 65 biotechnical drug products on the market. About 300 more biotechnical drug and vaccine products are in human clinical trials.

Biopharmaceuticals can be used to treat diseases such as anaemia, cystic fibrosis, dwarfism, hairy cell leukemia, hemophilia, and Kaposi's sarcoma. In gene therapy recombinant DNA technology is applied to correct single gene hereditary disorders. New types of vaccines are also being designed by applying transfected DNA expression of antigen molecules. Compared to many technologically mature industries the world pharmaceutical market is not consolidated to a significant degree. No single pharmaceutical company has a share of over 5 % of the $300 billion market. This would indicate that relatively free international competitive conditions prevail within the industry, but in individual markets related to particular medical conditions the market position of leading companies is often strong.

The drug compounds of today interact with only about 400 molecular targets in the human body. The total number of effective pharmaceutical compounds listed in Finland is about 900. These surprisingly low figures by themselves suggest that among the millions of chemical interactions within the human body there must be unexploited targets for new medical compounds. In 1998 only 35 new molecular entities (NMEs) were launched on the world market. Yet during that year alone altogether 11 000 pharmaceutical R&D projects were being carried out by the pharmaceutical industry. Over 3 600 projects had reached clinical phase.

Neurological therapeutic substances had a 13 % share of the R&D projects, cancer drugs 11 % and the share of potential antiinfective drugs was 9 %. The cost of developing new pharmaceuticals is high. In the United States, the Office of Technology Assessment estimated in 1993 that for drugs that first entered human testing in the period 1970—1982 the fully capitalised cost of developing a new pharmaceutical was $359 million in 1990 dollars. For drugs introduced in 1990, the cost was $500 million; one recent estimate is $400 million. It has been estimated that only one compound in 10 000 will succeed as a new pharmaceutical, and that the process takes approximately 12 years.

Efforts have been made to streamline the drug approval process while maintaining the high level of safety that has been achieved on the basis of existing regulations. In the United States The FDA Modernisation Act of 1997 was the first major legislative reform of food and medical product regulation in 35 years. A wider access to experimental drugs and medical devices was allowed by the reform, and filing and approval procedures of new therapies for serious or life-threatening conditions were streamlined.

Ultimately the high cost of medical research reflects a need to improve the methodologies of drug discovery and development. One reason for the high costs is the risk of discontinuities between in vitro and in vivo results. Similarly, preclinical animal models do not always eliminate the chance of negative results in clinical studies. The development of more accurate molecular, cell, and animal models is therefore a strategic priority for pharmaceutical R&D. Rapid advances in the Human Genome Project, drug screening methods, DNA chips, and computer modelling techniques are likely to increase the rate at which new original drug compounds are discovered.

The costs of pharmaceutical research could also be reduced if it could be determined at an early stage of research whether it is possible to develop suitable drug delivery methods for lead compounds that are interesting in other respects. The process of drug development is becoming increasingly systematic and computerised resulting in gradual cost savings. Subsets of human genes, such as thousands of those involved in cell-to-cell communication, can already be identified, and their proteins tested in high throughput systems to clarify their functions and discover potential medical applications. Such an approach was used recently to show that a compound, B1yS, has potential for stimulating the human immune y system against a variety of infectious diseases and strengthening the protective\impact of some vaccines.

Ideas for possible NMEs often first emerge in academic research carried out at universities. The ideas are taken up either by R&D companies or by members of the research teams who try to commercialise the new findings themselves. The overall result is an intricate network of co-operative arrangements sustaining wide-ranging research efforts even though only a small proportion of the projects lead to new commercially successful pharmaceuticals. Large pharmaceutical companies, as well as public and private financing institutions, have a rational incentive to spread the risks of R&D and support research on a large number of leads.

The human genome consists of about 80,000 genes stored in a total volume of three billion chemical base pairs of DNA. The Human Genome Project aims at completing an accurate sequence of the human genome by the end of the year 2003. At present the number of inherited diseases is estimated to be about 4 000. In the future the figure may be considered to be higher as hereditary cofactors may be recognised to be involved in diseases and conditions in which pathogen contagion, environmental factors, or even behavioural habits, play a dominant role. So far gene therapy methods have been developed for the treatment of a few singlegene disorders such as cystic fibrosis and severe combined immune deficiency.

In the future success may be achieved in developing treatments for polygenic diseases in which there have been mutations in several genes. It will be even more difficult to find treatments for multifactorial genetic disorders resulting from genetic factors and long-term environmental influences. Due to financing and human constraints science is at its weakest in solving puzzles that relate to slow changes that take place over decades. The complexity of, research tasks is increased by the existence of genetic polymorphisms, or slight genetic variations that influence the properties of expressed proteins in special circumstances, and by the fact that some genes have more than one function.

Fortunately technologies used in the identification of genes and in the clarification of their functions are developing rapidly. In practice these technologies are integrated when efforts are mounted to identify novel therapeutic treatments. For instance, the power of positional cloning techniques used in the study of family histories of disease inheritance is enhanced by the published results of the Human Genome Project, which relies on expressed sequence tag (EST) methodology. Similarly the discovery of genes coding for particular secreted proteins requires genetic mapping data and possibly computational sequence analyses.

The work to sort out the functions of genes lags behind the efforts to define partial gene sequences. There has been an intense debate on the wisdom of granting patents to gene sequences gained by EST methods, as the functions of such sequences are often not understood. Contrary to patent claims in other fields gene sequences as such disclose relatively little novel information, but patents obtained can hamper the use of the sequences in genetic probes and DNA arrays. The pace at which DNA chips and other technologies of functional genomics can be developed to some extent depends on the solution of patent issues.

A polymorphism can make its carrier more susceptible to some diseases, give unusual resistance to others, and even modify the efficacy of some pharmaceuticals. One example of an unusual resistance to a disease is a certain allele that obliterates the HIV-1 coreceptor on lymphoid cells and gives some resistance against HIV-1 infection. The allele may have developed approximately 700 years ago in Northern Europe, possibly as a result of the Black Death epidemic. Individuals who inherit this mutant gene from both parents are effectively devoid of the receptor and immune to an unusual degree to the HIV-1 infection.

Differences between genetic makeups of individuals can explain differences in reactions to some drugs. Only recently it was discovered that the reason why aspirin helps to protect some people from developing heart attacks is linked to the genetics of these individuals. A specific polymorphism related to their blood clotting platelets makes these people ten times more sensitive to the anticlotting effect of aspirin than individuals who do not have the polymorphism. It can be anticipated that the growing awareness of polymorphisms and the massive amounts of human genomic data available in the future will lead to focused drug development efforts that will take advantage of the particular genetic dispositions of various groups of patient populations.

During the last few decades the treatment of infectious diseases has relied on widespectrum antibiotic therapies. It has been suggested that this well established approach may change in the long run, when pathogen specific or at least narrowspectrum treatments are likely to become available. The need for such a change is based on the continuing proliferation of antibiotic resistant bacterial strains and challenges met in treating the patients suffering from immune system deficiencies. In the future new accurate and rapid diagnostic methods are likely to support the adoption of pathogen-specific treatments. Moreover, adjunctive immunotherapy strengthening the natural capabilities of the human immune system against specific pathogens may become possible in treating infectious diseases.

Pathogens are factors in more diseases than was suspected only a relatively short time ago. Pathogens play a leading role in dental caries (Streptococcus mutans) as well as in duodenal and gastric ulcers (Helicobacter pylori). Chlamydia pneumoniae may have a causative role in cardiovascular disease. Viruses and bacteria may also lie behind more cancers than is suspected at the moment. The World Health Organisation has estimated that a majority of some cancers are attributable to viruses, bacteria, or parasites. More than 1.5 million new cancer cases could be avoided each year by preventing the infectious disease associated with these diseases.

In the future it may become possible to identify the role of pathogens in conditions that are initially triggered by pathogen action, but then develop very slowly. Such conditions may include some autoimmune diseases, such as asthma, rheumatoid arthritis and multiple sclerosis. Future diagnostic methods may make it possible to discover remaining DNA fragments of the pathogens involved. As of yet we may even not know all existing types of infectious agents. An infectious agent of a new type, consisting of misfolded variations of the prion protein, has been discovered as the cause of the bovine spongiform encephalopathy (BSE) epidemic in cattle.

The misfolded protein is not eliminated by normal enzymes and accumulates in the brain causing a new form of CreutzfeldtJacob disease in humans and other fatal, but rare brain conditions. The discovery of new kinds of pathogens in the future cannot be precluded. For instance, it has been suggested that intra- and extracellular calcium phosphate deposits causing kidney stones and other medical conditions are produced by nanobacteria, that would be pathogens of a new type. The suggestion remains controversial. Even if eventually accepted, as in the case of the prion protein, it takes years for the scientific community to ascertain the validity a hypothesis that runs counter to well-established beliefs.

As pathogen species continue their evolution, new strains are likely to emerge in due course of time. The process could be facilitated by the fact that the immune systems of millions of people have been impaired by the AIDS epidemic. Immunosuppression resulting from treatments related to organ transplants has also become more common. Impaired immune function may allow the evolution of new pathogen variants that would have been eliminated by normal immune systems. Bacteria that normally live in close association with humans and are either neutral or beneficial may be transformed to become malevolent by mutations induced by natural evolution, chemicals, pharmaceuticals or other pathogens.

Acinetobacter and Xanthomonas were originally innocuous, but in a few years they evolved into multidrug-resistant strains that are capable of causing potentially fatal blood borne infections in hospitalised patients. Also it is possible that in the future more pathogens may cross over from animals to afflict humans. AIDS originated from chimpanzees and BSE-CDJ from cattle. In Central Africa there have been epidemics of the Ebola and Marienburg virus while in China there have been lethal cases of bird influensa.

Lyme disease is one example of a human disease transmitted by animals in industrialised countries. In the future xenotransplants may be used to save the lives of thousands of people, but the method also could increase the risk of introducing novel infectious agents of animal origin into the human population. Drug resistance is a manifestation of the ability of bacteria and viruses to evolve and adapt to changing circumstances. Several bacterial strains have developed multidrug resistance. They include Streptococcus pneumoniae, Enterococcus faecaCis, Mycobacterium tuberculosis, Pseudomonas aeruginosa and Staphylococcus aureus as well as several strains of Listeria.

The stockpile of efficacious antibiotics of the `first generation' has been depleting at a faster rate than new medicines have been developed. Resistant strains have emerged and spread at a surprising speed all over the world. Bacteria use a variety of mechanisms to fight antibiotics. Some produce enzymes that render first generation antibiotics such as penicillin harmless. Sometimes bacteria develop pumping or efflux mechanisms to remove antibiotics from the cell. Bacteria can also change the shape of molecules that are targeted by an antibiotic.

One of the resistance mechanisms against tetracycline is a pumping system that removes not only tetracycline but also other antibiotics from the bacterial cell. In another method the tetracycline target, proteinproducing ribosome, is modified and the structure on which the drug binds is changed. The third mechanism involves the modification of the tetracycline compound itself, but it has been discovered only in two anaerobic bacterial strains. Genetic information coding the resistance mechanisms is usually stored on plasmids, tiny loops of DNA.

Many plasmids have originally evolved to help bacteria survive hasards in the environment. Plasmids are relatively easily transmitted between individual bacteria of the same species and even between individuals of different species. Resistance genes can also be transferred by viruses from one bacterial cell to another. Occasionally, a bacterium can also acquire a gene from the remains of a dead bacterial cell. Compounds that would neutralise resistance mechanisms would be useful in developing `antibiotics of the second generation' or adjuncts that would enhance the efficacy of existing antibiotics. For instance, the efflux pump would be a tempting target for countermeasures against multidrug-resistant bacteria.

Of the known efflux mechanisms one of the most efficient is that based on the multidrug transporter protein MdfA. By using a single negatively charged amino acid residue the transporter protein is able to identify a number of antibiotic compounds and remove them from the bacterial cell. However, as the various resistance mechanisms are based on systems that have emerged in natural evolution, they are likely to be well protected against interference. The ability of these systems to evolve is already proven by the drug resistance itself. That is why the search for antibiotics of the second generation is not likely to be easy.

Usually investments in research and development expand technical opportunities and create opportunities for creating new products. However, new technologies can have negative effects as well, because they can reduce the potential for innovation in neighbouring fields of technology. In biotechnology bacterial resistance due to the use of antibiotics as growth-promoting factors in animal husbandry may eliminate opportunities that could be used to develop pharmaceuticals for humans. A case in point is everninomicin, which has shown some promise as a potential new antibiotic; it contains compounds that have not been applied in any class of medicine used in people.

However, everninomicin resembles another oligosaccharide, avilamycin, that has been used in animal husbandry for several years. In a recent study it was found that, as a result of the use of avilamycin, there already are bacteria whose susceptibility to everninomicin has declined even before the human use of the antibiotic has started. It is evident that antibiotic resistance phenomena can significantly increase the costs of antibiotic drug research and reduce the profitability of future drug development efforts. Outer membrane proteins of several pathogens are obvious targets for future drug development efforts, as some of the most successful antibiotics, including penicillin and vancomycin, impede the proper synthesis of the bacterial cell wall.

Chlamydial infections are common, and the major outer membrane protein of Chlamydia trachomatis is of great interest in vaccine development. However this protein exhibits considerable antigenic variation, as do many other focii of vaccine research. The cell envelope of Mycobacterium tuberculosis contains several unique polymers and the inhibition of their biosynthesis is one aim in efforts to develop narrow-spectrum pharmaceuticals for the treatment of tuberculosis. Some bacterial species communicate. They send chemical messages by secreting messenger compounds. A common household bacterium Pseudomonas aeruginosa, responsible for many hospital infections and a leading cause of death in cystic fibrosis patients, resists antibiotics by forming a sticky protective biofilm.

Its formation is triggered by a homoserine lactone chemical that the bacterium releases. When there is enough of the chemical in the environment, the produsction of biofilm is started. A gene controlling the process has been discovered. Possible targets for future antimicrobial compounds include the messenger chemical, the gene controlling biofilm formation, and intermediate steps in the process. When the formation of biofilm is prevented, Pseudomonas is left almost defenceless against antibiotics. The development of antiviral drugs and vaccines against human immunodeficiency virus (HIV) and hepatitis C virus (HCV) has proved very difficult, as these viruses mutate constantly.

Each HIV virion carries two complete genomic strands based on RNA. Natural recombination of these strands is possible when a single cell is infected with two different. strains. Resulting recombinant HIV strains have been found in patients in regions where multiple HIV variants circulate. Variations arising from recombination poses difficult problems for the development of HIV drugs and vaccines. In the future it may become possible to design therapeutics that would attack specific vital, slowly mutating RNA sections of the HIV.

Therapeutics focusing on RNA could utilise ribozymes, enzymes that can be modified to bind with any part of the RNA molecule. Ribozymes could be administered as drugs or, in a more distant future, a gene therapy could be used to insert a ribozyme encoding gene into host cells. Added protection against the emergence resistant strains could be achieved by targeting multiple sequences of the viral RNA. In experimental studies with HIV success in the form of very high inhibition rates of viral progeny production has been reported. Immunisation is the most cost-effective way of dealing with pathogens, and in the future DNA vaccines are likely to be developed against a number of human diseases.

The mechanism used in DNA vaccines resembles viral infection. The aim is to use the biosynthetic machinery of the host cell to synthesise a foreign protein. The protein then becomes an antigen to which the immune system reacts resulting in immunity. Plasmid DNA coding a target protein, for instance a protein found on the surface envelope of a pathogen, is introduced into the host organism by injection or particle bombardment. A small number of host organism cells absorb the DNA code and express the protein.

The host immune system then reacts to the protein and the result is protection against the pathogen as- the immune system will be able to recognise the antigen protein in the future and will produce antibodies rapidly. It may become possible to develop DNA vaccines that contain not just one but even thousands of DNA strings expressing proteins of a particular pathogen. Such an approach might be used in future emergencies, when a vaccine is urgently required against a new virulent pathogen. DNA vaccines are evolving into a versatile methodology. Trials of an HIV vaccine were started in 1999 in Finland.

Elsewhere research is being carried out on the use of DNA vaccines in inhibiting symptoms of allergic asthma as well as conferring protection against Hepatitis B disease and malaria. Another vaccine is being developed against tick-born encephalitis caused by flavivirus. In many of the studies the DNA vaccine is being administered together with an immune system adjuvant in order to elicit a stronger immune system reaction and better protection. DNA immunisation has also been used to prevent or inhibit tumour development. Clinical trials have been carried out on patients with lymphoma and metastatic melanoma.

Tumorigenesis

Tumorigenesis can be best characterised as a process caused by five or six accumulating genetic alterations. Mutations in oncogenes promote proliferation of cancer cells, while changes in genes governing programmed cell death (apoptosis) inhibit the onset of a natural cell elimination process. Mutations in apoptosis genes such as P53 and in oncogenes such K-ras are often implicated in studies of tumorigenesis. In the future DNA tests may be used in screening large populations for cancers enabling early detection and treatment. After diagnosis the genetic profile of a patient's primary tumour could be used to define prognosis and optimise treatment.

Most present day cancer treatments involve the use of relatively blunt methods such as surgery, poisonous chemicals and radiation. In the future it may become possible to develop increasingly smart methodologies to diagnose not only the type of cancer, but also its stage and characteristics. As all cancers are genetically unstable, lots of genetic variation and mutations can be expected to occur between various stages and even regions of a single tumour. That is why treatments relying on a single antibody or a single DNA target are not likely to be very successful. Gene therapy relying on several genes may be a part of the solution.

Over the last few years more than one hundred clinical trials involving the use of gene therapy in cancer treatment have been approved in the world. Viral vectors that have been tested include retroviruses and adenoviruses. Nonviral methods include liposomes and direct injection. In targeted delivery tumour- and tissue-specific antigens have been the selected binding sites. Therapeutic genes used in the trials have included cell suicide genes, genes preventing the functioning of oncogenes (based on antisense technology and ribozymes) and genes coding enzymes that enhance the efficacy of cancer drugs. Treatment of many cancers is complicated by the physical structure of solid tumours.

Cancer cells degenerate rapidly, and solid tumours always have some dead tissue that protects the tumour against the penetration of drugs and vectors carrying therapeutic genes. Solid tumours also hamper adequate blood supply to adjacent healthy tissues, and as a result neighbouring cells as well as cancer cells themselves often suffer from lack of oxygen. Hypoxic cells are usually resistant to radiotherapy and chemotherapy. However it may become possible to use hypoxia response elements as triggers in activating therapeutic genes in healthy tissues adjacent to the tumour. Rapidly growing tumour cells fight for oxygen by expanding the network of blood vessels feeding them.

It may become possible to prevent the growth of vasculature serving the tumour cells, and to destroy selectively these blood vessels, thus starving and suffocating the cancer to death. Promising results have been achieved in animal trials in which antiangiogenic factors have been tested in restricting blood supply to tumours. Tumour blood vessels can be selected for drug targets, because the endothelial cells of rapidly growing blood vessels contain growth factors that are lacking from normal veins. Vascular endothelial growth factor seems to be elevated in many cancer patients regardless of the histological type of the cancer. In some cases the human immune system is able to eliminate cancer cells before they become numerous enough to be a threat.

Indirect evidence is presented by a particular immunity mechanism, that has serious side effects. About one in every 1000 women having breast or ovarian cancer develops a protective immunity, in which the immune system targets a cancer cell protein known as cdr2. Unfortunately the same protein can also be found in the cerebellum area of the brain. Consequently the immune response also affects the brain, and these women develop a neurological condition known as paraneoplastic cerebellar disorder that is usually diagnosed before the cancer.

Cytotoxic T lymphocytes (CTIs) play an important role in immunity and resistance against cancer. Antigens released by dying cancer cells are carried by the blood stream to dendritic cells, which then alert CTLs to attack cancer cells exhibiting the antigens. Animals studies have shown some success in fighting cancer by vaccinating animals with dendritic cells that have been made to carry tumor antigens. In the future it may become possible to mobilise the strengths of the human immune system against cancer cells of various types.

Particular groups of people, identified in population screenings as having a high risk of developing a cancer of a certain type, could be offered some protection with the help of DNA vaccines targeting some features of cancer cells. Moreover, it may become possible to define mechanisms that protect cancer cells from being discovered by the immune system. If it became possible to turn off the protection, the immune system would be able to destroy the malignant cells.

For the pharmaceutical industry human and pathogen enzymes are of immense importance. Aspirin, an analgesic that has dominated the market for a century, blocks the production of the cyclo-oxygenase (COX) enzyme thus inhibiting the synthesis of prostaglandins. These fatty acids help the body to produce symptoms of pain and inflammation. Unfortunately the human stomach lining is protected by the COX enzyme and its inhibition easily leads to stomach ulcers. The discovery of the COX-2 enzyme provided the drug industry with a new target that helps to avoid gastrotoxic effects. The annual world market for the new COX-2 inhibitors is estimated to amount to $5 billion.

The influenza virus uses an enzyme called neuraminidase to free itself from the human cell where it was synthesised. Recently neuraminidase inhibitors have been developed, and it has been suggested that inhaling these medicines daily during an influensa outbreak could reduce the chance of catching the disease. The drug appeared to shorten bouts of flu by about 2 days in trials in Europe and Australia, but a similar study among U.S. patients produced lesser results.

Enzymes repairing DNA are of great interest to the pharmaceutical industry. They sustain life, protect against cancer, remedy damages caused by ultraviolet (UV) light and may prevent some of the biological changes associated with ageing. Treatments based on DNA-repairing enzymes could be particularly useful in preventing skin cancer, one of the most rapidly increasing forms of cancer. In humans an enzyme called glutathione peroxidase counteracts harmful effects of UV radiation. It seems that a virus causing a skin disease has been able to acquire the gene coding the enzyme; as a result the virus has gained some protection against UV radiation.

Recently it has been shown, that an error prone repair enzyme may be responsible for much of DNA damage associated with ultraviolet light. An enzyme called telomerase, not found in normal cells, repairs telomeres, specialised nucleic acid structures located at both ends of linear eukaryotic chromosomes. Each time a cell divides, telomeres get shortened, and finally, when telomeres get very short, cell division is prevented. The mechanism may act as a molecular clock marking and controlling ageing processes. In some cancer cells telomeres are maintained by telomerase activity, enabling uncontrolled growth.

In the future telomerase may be used to counteract some features of the ageing process. Similarly substances inhibiting telomerase in cancer cells may be used in the development of new cancer treatments. A special case of the power of DNA repairing enzymes is Deinococcus radiodurans, a microbe discovered in 1956 in a can of meat that had been radiated in order to sterilise it. The microbe not only endures acute radiation doses of up to three million rads, but it grows even when exposed to radiation continuously. The bacterium is able to repair damage to its DNA efficiently. The trait probably evolved as a response to extreme droughts that damaged the bacterium's DNA.

DNA polymerases carry out a wide variety of synthetic tasks during DNA replication and repair. These enzymes range from variants that deal with single nucleotide gaps to enzymes repairing gaps of up to kilobase size. Ribozymes are special enzymes capable of cleaving and repairing RNA at specific sites. It is possible that in the future combinations of new approaches such as gene therapy, ribozymes and immunotherapies will be used simultaneously to inhibit multiple stages of the life cycles of infectious agents.

Enzymes called phosphoinositide kinases are located on the cell membranes of most types of cells; they receive and relay chemical signals. Some types of disorders seem to be related to the signalling system, causing cells to proliferate and travel to distant parts of the body, leading to cancers. The signalling system is also important for nerve cells. In a post-mortem study of depressed suicide victims the phosphoinositide system was found to be impaired by about 30 % compared to the normal level.

Pharmaceuticals for the Central Nervous System

We may recall how attempts to explain gastric ulcers on the basis of personality and behaviours of patients were not very useful in developing medical treatments to the condition. Similarly efforts to explain medical conditions related to the central nervous system on the basis of personality traits, personal experiences or behaviours have not been very useful in the development of efficacious pharmaceuticals. Instead, psychopharmaceutical drug leads have been often discovered on the basis of random events. The value of monoamine oxidase inhibitors in treating depression was discovered in the 1950s as a result of a trial of a drug that was designed for the treatment of tuberculosis.

The efficacy of lithium in dampening both the manic and depressive phases of the bipolar disorder has been known for half a century, but a comprehensive explanation was found only recently. It seems that lithium stabilises glutamate uptake in presynaptic nerve endings in the cerebral cortex. Glutamate is an important excitatory neurotransmitter found in most regions of the brain. The discovery of neurotransmitters and their receptors put the psychopharmacological drug discovery process on a rational footing. The effects of chemicals that relay signals between neurons can be amplified, reduced, mimicked and blocked.

As a result new pharmaceuticals have been developed, and researchers have improved their understanding on how biomolecular processes in the central nervous system are related to mood disorders and addictions. It has also been shown, how abuse or neglect of young mammals and humans may lead to biomolecular changes in the central nervous system. Restoring the biomolecular balance could help these individuals to overcome the effects of past events. The regulation of neurotransmitters such as serotonin, dopamine, norepinephrine and acetylcholine is the aim of many ongoing drug research efforts. Each of these research areas is promising in itself, but the central nervous system is a complex environment, in which neurotransmitters and receptors have several variants and different functions in separate circuits.

A pharmaceutical relating to a single transmitter may provide a step into the right direction, but may not be enough to restore the overall balance of the complex system. Therefore it is likely that cocktails of transmitter related medicaments will be designed in the future for the treatment of various subtypes of mood and addictive disorders. Restoring the balance between several transmitter systems may be the key in the treatment of, for instance, panic disorder. It is likely that pharmaceuticals will be developed in the future for the treatment of some forms of obesity and substance abuse. Some of the dopamine receptors of the brain are involved in addictive behaviours, indicating that medications facilitating changes in these behaviours may be possible in the future.

Similarly, if sensory reactions caused by food are particularly strong among some obese people, in the future these reactions may be alleviated with appropriate medication. Antiobesity agents in clinical trials include leptin, a compound discovered in 1994. High leptin levels seem to signal to the central nervous system that energy reserves are sufficient, whereas low levels signal the need for increased energy intake. Molecular mechanisms of several neurodegenerative diseases have been discovered in the 1990s. As the causes of the diseases are now better understood, new subtypes have been discovered, and the classification principles of the diseases can be redefined. Mutations in the so-called tau protein seem to lead to most cases of Parkinson's disease, while alterations in the a-synuclein protein are the main cause of Alzheimer's disease.

In the future early diagnosis methods may become available enabling pharmaceutical therapies that will at least delay the onset of typical symptoms. Efforts to develop drugs against neurodegenerative diseases may lead to the emergence of compounds that have some relevance in enhancing individuals' capacity to learn, remember, and recover mentally after stress. Cholinergic and glutamatergic neurotransmission mechanisms participate in learning and memory processes, and they are important target areas in research on pharmaceutical leads that would alleviate the symptoms of Alzheimer's disease.

Similar compounds could be beneficial also for children having certain types of learning disabilities. It is likely that mental conditions such as schizophrenia comprise a great number of variants and causes. Schizophrenic symptoms arising in adolescence or adulthood can be a result of perturbations that have taken place in corticolimbic circuitry during prenatal development or in early childhood. New diagnostic tools would be urgently required to enable accurate diagnosis of subtle variants of schizophrenia and other conditions affecting the central nervous system.

In the case of depression, the level of corticotropin-releasing factor (CRF) would seem to be an important target for measurements. Some cases of mental disorders are caused by infectious agents such as herpes virus and Boma disease virus. Accurate diagnostic tools of the future may create an improved basis for the treatment of various subtypes of medical conditions involving the central nervous system.

Drug Manufacturing Technologies

Biotechnology creates methods that complement traditional techniques of pharmaceutical production and others that can revolutionise production techniques and reduce manufacturing costs significantly. Chiral compounds produced by using biocatalysts provide an example of the use of biotechnology in enhancing traditional production techniques. All proteins in the body are stereospecific; that is, they can react differently to the two chiral forms of molecules. One chiral form or enantiomer of thalidomide is a useful pharmaceutical, while the other causes birth defects.

In the early 1990s there were about 500 synthetic chiral drugs on the market, but about 90 % of them were sold in a form that included both enantiomers. Since then pharmaceutical companies have made great efforts to replace the mixtures with single enantiomers. By using enzymes drug manufacturers can avoid the production of an unwanted enantiomer and reduce harmful side effects. Biotechnology enables the recruitment of bacteria, plants and animals for pharmaceutical production. Procaryotic cells (bacteria) or eucariotic cells (yeasts, animals) can be used in the production of proteins of therapeutic interest, but not all cell types can be harnessed to produce any protein. For instance, bacteria cannot produce glycoproteins, since they lack the capacity to glycosylate.

The use of Gram-negative bacteria is hampered by toxic lipopolysaccharide contained in their outer membranes that are more complex than those of Grampositive species. Among plants tobacco is a favoured species used as a host for genes coding pharmaceutical proteins. Tobacco is relatively easy to modify, grows rapidly, and produces large amounts of seed, facilitating production increases when necessary. Transgenic plants seem to be particularly suited for the production of fully functional antibodies, including complicated and expensive secreted IgA antibodies for anti-inflammatory therapeutics.

The problems of using transgenic plants in pharmaceutical production include low yields and difficulties in extraction and purification. A remedy is to make plants secrete the desired proteins so that they can be harvested continuously. Tobacco plant roots have already been experimentally engineered to secrete xylanase enzyme into a hydroponic medium from which the enzyme can be easily extracted. The ability of roots to secrete proteins originally evolved to protect the plants against bacteria and other pathogens. Another secretion pathway that may be exploited in the future is leaf guttation, commonly known as morning dew.

Even though transgenic plants can be made to produce many human proteins, the plants are not always able to carry out the necessary postsynthetic modifications required to make these proteins fully functional. In transgenic animals the modifications are often performed as a matter of course. That is why transgenic animals are likely to be a preferred source of many pharmaceutical proteins, such as hormones, enzymes and vaccine proteins. In the future human lactoferrin for the treatment of gastrointestinal infections as well as various expensive blood clotting factors are likely to be produced at a relatively low cost in the milk of transgenic cows, pigs or goats.

Drug Delivery Techniques

In drug development projects problems related to pharmacokinetics and drug delivery often cause considerable costs and risks. Methods that can be used to discover and solve potential problems at an early stage of drug development would be important for reducing the costs of drug development. Drug delivery to the central nervous system is a particularly tough challenge. The blood-brain barrier consists of unusually tight capillary walls; their protective effect is enhanced by glial cells wrapped around them.

The barrier prevents the entrance of large molecules, including proteins, biopharmaceuticals and water soluble chemicals. Among the few peptides passing through the barrier are insulin and leptin, and most of the lipid soluble molecules can also pass through. Injectable biodegradable microcapsules and microparticles releasing drug compounds in a controlled manner as well as collagen gels are some of the options for reaching the central nervous system. In the long term it may become possible to engineer and inject modified stem cells that produce the desired protein within the central nervous system. Such cells could offer the additional benefit of replacing dysfunctional neurons, thus strengthening weakened brain functions. New general drug delivery new methods such as transdermal and nasal routes as well as aerosols penetrating deep into lungs are likely to be used increasingly in the future.

From the point of view of biopharmaceuticals pulmonary delivery is an interesting option, as the route is likely to prove particularly suitable for the delivery of peptides and protein based compounds, including insulin. In the future transgenic pharmaceutical production technologies may be combined with genes that code structures that give pharmacokinetic protection to the drug compounds. Encapsulation in polylactic acid materials has already been approved for use in controlled drug delivery. In a method patented in the United States, drugs can be incorporated into macrophage cells. The cells can then be injected intravenously, and the final phases of delivery take advantage of the homing mechanism of the macrophages.

2

Biocatalysis

Many reservations with respect to biocatalysis have been voiced over the last 20 years. The critics, often focusing on the drawbacks, have contended that

— enzymes only feature limited substrate specificity,

— there is only limited availability of enzymes, only a limited number of enzymes exist,

— protein catalyst stability is limited,

— enzyme reactions are saddled with limited space-time yield, and

— enzymes require complicated co-substrates such as cofactors.

The renaissance of biocatalysis, supported by the advent of recombinant DNA, is only about 20 years old. Recently, several publications have appeared which deal specifically with the attitudes listed above. Most of the points above can either be refuted or they can be levied against any novel catalytic technology; the situation with some competing technologies such as chiral homogeneous catalysts is similar to that with enzymes.

Enzymes only feature limited substrate specificity: Often, enzymes designed to convert small molecules such as hydrogen peroxide, urea, fumaric acid, or laspartic acid feature extremely narrow substrate specificity; the corresponding enzymes catalase, urease, fumarase or aspartase, and l-aspartatedecarboxylase take either few other substrates, such as alkyl peroxides in the case of catalase, or no other substrate, such as urease, which only converts urea.

On the other hand, very large enzymes acting as multi-enzyme complexes such as nonribosomal peptide synthetases (NRPSs) are often highly specific. Ordinary-sized enzymes working on medium-sized substrates, however, in most cases feature broad substrate specificity, a fact already noted by Rasor and Voss.

There is only limited availability of enzymes. Until very recently, limited availability of enzymes was indeed a major problem. About ten years ago, with 3196 different enzymes already listed in Enzyme Nomenclature, only about 50 enzymes were fully characterised and only about a dozen enzymes available commercially on a regular basis. However, recombinant DNA technology, discovered in 1978 by Cohen and Boyer, over the next 20 years allowed enzymes to be produced much more efficiently, in higher purity, and more inexpensively, so that today a multitude of enzymes are available not only from established suppliers such as Sigma-Aldrich-Fluka (Milwaukee, WI, USA), E. Merck (Darmstadt, Germany), Mercian (Tokyo, Japan), or Roche Diagnostics (Mannheim, Germany) but increasingly also from smaller, more focused suppliers such as Biocatalytics (Pasadena, CA, USA) or Jülich Fine Chemicals (Jülich, Germany). The argument of unavailability or scarcity will be less and less justified in the future.

Only a limited number of enzymes exist. This criticism, while depending on the observer's position, is indeed a drawback at the moment. Although enzymes have been found for every conceivable organic chemical reaction except the hetero-Cope rearrangement, there are enzymes sought for many more reactions than there are enzymes available. If enzymes were inferior catalysts this situation would not arise, of course. In fact, enzymes are often superior catalysts, so superior is fact, that the community seeks plenty more of them.

Protein catalyst stability is limited. This is one of major drawbacks of enzymes. They commonly require temperatures around ambient to perform (15–50°C), pH values around neutral (pH 5–9), and aqueous media. In addition, any number of system components or features such as salts, inhibitors, liquid–gas or liquid– solid interfaces, or mechanical stress can slow down or deactivate enzymes. Under almost any condition, native proteins, with their Gibbs free enthalpy of stability of just a few kilojoules per mole, are never far away from instability.

Enzyme reactions are saddled with limited space-time yield. The notion that biocatalysts are slow catalysts is false. Slow catalysts, applied at low concentrations, certainly lead to low space-time yields. However, optimised syntheses not only produce very good selectivities or total turnover numbers but also satisfactory to excellent space-time yields. Examples with such good s.t.y. values are - in commodity biochemicals, the synthesis of l-aspartate from fumaric acid and ammonia with space-time-yields of up to 60 kg $(L \cdot d)^{-1}$, and in advanced pharmaceutical intermediates, kinetically-controlled peptide synthesis to kyotorphin (Tyr-Arg) catalysed by α-chymotrypsin from maleyl-l-Tyr-OEt and Arg-OEt, employing a highly soluble protecting group at the electrophile. Space-time-yields of 1.34 kg $(L \cdot d)^{-1}$ have been achieved. The question of high volumetric productivity is coupled to the solubility of substrates. High space-time-yields have been demonstrated to be correlated with high solubilities of substrates.

Enzymes require complicated co-substrates such as cofactors. Much has been made of the requirement of some enzymes for cofactors, such as nicotinamide-containing compounds, NAD(P)(H), for dehydrogenases; flavin compounds, FMN or FAD, for oxidases; pyridoxyl-phosphate, PLP, for transaminases and decarboxylases; thiamine pyrophosphate, TPP, for carboligases, and vitamin B12 for glycerate dehydratase, among others.

The scale-up of cdecarboxylation to l-alanine with the help of PLP-requiring l-aspartate decarboxylase, or of reductive amination of trimethylpyruvate to l-tert-leucine with the help of NADHrequiring leucine dehydrogenase demonstrates the feasibility of industrial processing with cofactor-requiring enzymes. The implementation also gives credence to the suggestion that cofactors are no longer the dominating cost component, as was believed until recently. Requirements for cofactors constitute a technological challenge but one that has been met successfully and so should not be regarded as impeding the use of biocatalysts in processing.

Advantages of Biocatalysts

The biggest advantage of enzymes is their often unsurpassed selectivity. While enzymes are used beneficially to increase chemical selectivity or regioselectivity of a reaction, their biggest advantage lies in the differentiation between enantiomeric substrates, a pair of substrates with Gibbs free enthalpy differences between the Rand the S-enantiomer.GRS of around 1-3 kJ mol^{-1}. With enzymes, enantioselectivities of > 99% e.e. can be achieved routinely, although by no means in every case. This fact becomes increasingly important for using biocatalysts in the synthesis of advanced pharmaceutical intermediates, as regulatory agencies require separate toxicological studies for every impurity comprising above 1% of the content.

The fact that enzymes are active mostly at mild, near-ambient conditions of temperature and pH and preferentially in aqueous media is often regarded as an advantage rather than a drawback nowadays. Goals for industrial processing such as "sustainable development", "green chemistry", or "environmentally benign manufacturing", an increasingly important boundary condition for industrial activity in a large part of the world, would be much harder to attain without the availability of biocatalysts which tolerate and require such conditions. Biocatalysts are able to catalyse an increasing breadth of reactions. This breadth translates into an increasing number of applications of biocatalysts on and industrial scale. Increasingly, biocatalysts are combined with chemical catalysts or utilised in a network of reactions in the cell.

Drawbacks of Biocatalysts

There are three essential drawbacks of today's biocatalysts:

1. biocatalysts are often not sufficiently stable in the desired media,
2. too few biocatalysts exist for the desired reactions from available substrates to targeted products, and
3. development cycles are too long for new and improved biocatalysts.

Biocatalysts are often not sufficiently stable in the desired media. As mentioned above, this still is an essential drawback of biocatalysts. As even conformational changes of less than an Ångstrom can cause a precipitous decline in activity, retention of activity is a stringent criterion for the integrity of a protein molecule. Enzymes deactivate under a range of conditions such as extremes of temperature or pH value, physical forces such as cavitation by pumps and aqueous -organic or gas-liquid interfaces, or specific covalent interactions.

Too few biocatalysts exist for the desired reactions from available substrates to targeted products. This argument can be approached from two vantage points. There are biocatalysts for almost any reaction. However, most biocatalysts (there are now more than 4000 known) are either not well characterised, or proprietary, or at least not commercially available. The situation, though, improves steadily: just ten years ago, only about a dozen enzymes were available commercially, whereas nowadays the number has increased more than tenfold. Further rapid progress is to be expected.

Development cycles are too long for new and improved biocatalysts. The development of some flagship biocatalytic processes of today took between 10 and 20 years: development of the acrylamide process took 20 years, of the Lonza process to l-carnithine 15 years. Compared to patent lifetimes around the same length, such durations can easily be deemed far too long for many applications. One reason for such timelines is the as-yet incomplete knowledge base of biotechnology and biocatalysis.

With an improved knowledge base stemming from intensive research efforts, development times will certainly be decreased. Shortening the development cycle time for biocatalyts should therefore be a topic of active research. It is. interesting, however, to pose the question whether long development times are particular for biocatalysis. Even a cursory inspection of the situation of homogeneous chiral chemical catalysis reveals that the situation is no different than for biocatalysis..

Characteristics of Biocatalysis

From different disciplines, biotechnology and biocatalysis are seen from very different angles and perspectives (Figure 1). Chemistry and chemists emphasise a molecularly-oriented perspective dominated by compounds and transformations, whereas chemical engineering and thus chemical engineers favor a process-oriented perspective of reactions and processes; lastly, biology and its practitioners contribute a systems-oriented perspective of description at the organism level as well as in their view of evolution.

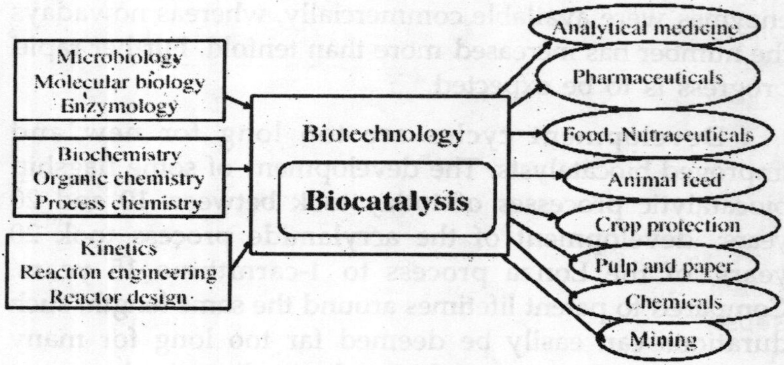

Figure 1. Central role of biocatalysis and biotechnology between interdisciplinary feeder sciences and multiple user industries.

Different parts of each of the three disciplines are needed for the successful practice of biocatalysis: biochemistry and organic chemistry from chemistry; molecular biology, enzymology, and protein biochemistry from biology; and catalysis, transport phenomena, and reaction engineering from chemical engineering are indispensable. Both biotechnology and biocatalysis are interdisciplinary areas; as most practitioners tend to hail from one of the three major contributing disciplines, hardly anybody has an equally strong command of all the sub-disciplines of biocatalysis.

There are not only many contributing disciplines for biotechnology and biocatalysis, but also many areas of application:

— production and transformation of compounds, mainly in the chemical and pharmaceutical industry,

— analytics and diagnostics, mainly in medicine, and

— environmental protection and bioremediation (reconstruction of the environment).

The areas of application differ from the industries which apply them; the most important ones are the pharmaceuticals, food, fine chemicals, basic chemicals, pulp and paper, agriculture, medicine, energy production, and mining industries (Figure 1).

Biocatalytic Transformations

Biotechnological transformations include a broad range of processes, ordered according to the number of biologically performed process steps and the complexity of the substrates (Figure 2):.

— Fermentations transform raw materials such as sugar, starch or methanol, often in industrial mixtures such as molasses or corn sleep liquor as carbon sources with living cells to more complex target products.

— Precursor fermentations start with defined educts and transform these, again with living cells, to the desired target products.

— Biotransformations transform defined precursors in a series of defined (not always known) steps with enzymes or resting cells to a desired target product.

— In enzyme catalysis, frequently crude extracts or partially purified enzymes, which only have to be free of side activities, are utilised for the transformation from defined substrate to target product.

— Purified enzymes are rarely used for the production of chemicals, possibly only for the production of highly priced fine chemicals such as pharmaceutical actives.

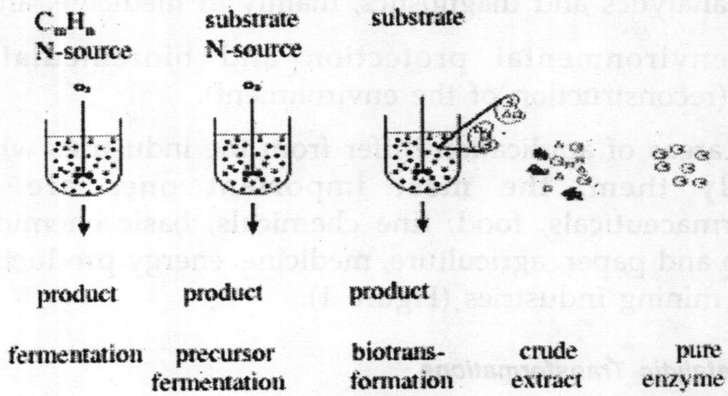

Figure 2. Biocatalysis as a continuum between fermentation and transformations with pure enzymes.

The limits between the areas are blurred: biotransformations and enzyme catalyses with crude extracts or pure enzymes are often summarised under the term "biocatalysis". "Biocatalytic processes" are taken to mean transformations of a defined substrate to a defined target product with one or several enzyme-catalyzed steps. Biotechnology and biocatalysis not only differ from conventional processes by featuring a different type of catalyst but they constitute a new technology base:

— Whereas the raw materials base of a conventional process is built on coal or oil, the base for biotechnology is sugar, lignin, or animal or plant wastes.

— Whereas conventional processes are dominated by unit operations such as distillation, often at high pressure or temperature, in biotechnology unit operations, such as membrane processes, chromatography or biocatalysis are prevalent.

> Whereas conventional products are often achiral organic molecules or polymers, the product range of biotechnological processes often encompasses chiral molecules or biopolymers such as proteins, nucleic acids, or carbohydrates.

Compared with other kinds of catalysts, for example homogeneous catalysts, in which ligands are responsible for specificity, and heterogeneous catalysts, in which catalytically active centers are attached to solid carriers such as zeolites or metal oxides, enzymes feature the advantages and disadvantages listed in Table 1.

Table 1. Advantages and disadvantages of biocatalysts and enzymes.

Advantages	Disadvantages
Very high enantioselectivity	often low specific activity
Very high regioselectivity	instability at extremes of T and pH
Transformation under mild conditions	availability for selected reactions only
Solvent often water	long development times for new enzymes

Whereas enzymes often feature great advantages in terms of selectivity, their stability is often insufficient. Additionally, long development times of new biocatalysts owing to an insufficient knowledge base of biocatalysis and biotechnology remain a problem and a challenge.

Applications of Biocatalysis

Owing to two very strong and important driving forces the chemical industry of the future will look considerably different from today's version: cost and margin pressure resulting from competition in an increasingly open market-oriented economy, and operation of the industry in a societal framework which puts emphasis on a clean (or at least less polluted) environment. Processing with a view towards this new set of conditions focuses on the development of production routes with fewer processing steps, with higher yields on each step, to save material and energy costs.

Less waste is generated, and treatment and disposal costs go down. Both pressures come together in the cases of environmental compliance costs. In many cases, such as high-fructose corn syrup, or biotechnology and biocatalysis offer technology options and solutions that are not available through any other technology; in such situations such as acrylamide, nicotinamide or intermediates for antibiotics, biotechnology and biocatalysis act as "enabling technologies".

Green chemistry is an overarching approach that is applicable to all aspects of chemistry. Green chemistry methodologies can be viewed through the framework of the "Twelve Principles of Green Chemistry":

1. It is better to prevent waste than to treat or clean up waste after it is formed.
2. Synthetic methods should be designed to maximise the incorporation of all materials used in the process into the final product.
3. Wherever practicable, synthetic methodologies should be designed to use and generate substances that possess little or no toxicity towards human health and the environment.
4. Chemical products should be designed to preserve efficacy of function while reducing toxicity.
5. The use of auxiliary substances (e.g., solvents, separation agents, etc.) should be made unnecessary wherever possible, and should be innocuous when used.

6. Energy requirements should be recognised for their environmental and economic impacts and should be minimised. Synthetic methods should be conducted at ambient temperatures and pressures.
7. A raw material or feedstock should be renewable rather than depleting wherever technically and economically practicable.
8. Unnecessary derivatisation (blocking group, protection/deprotection, temporary modification of physical/chemical processes) should be avoided wherever possible.
9. Catalytic reagents (as selective as possible) are superior to stoichiometric reagents.
10. Chemical products should be designed so that at the end of their function they do not persist in the environment and they do break down into innocuous degradation products.
11. Analytical methodologies need to be further developed to allow for real-time, in-process monitoring and control prior to the formation of hazardous substances.
12. Substances and the form of a substance used in a chemical process should be selected so as to minimise the potential for chemical accidents, including releases, explosions, and fires.

Catalysis offers numerous advantages for achieving green chemistry goals: novel, high-yield, shorter process routes; increased selectivity; and lower temperatures and pressures. Biocatalysis combines the goals of all three topics above. Biocatalysts, as well as many of the raw materials, especially those for fermentations, are themselves completely renewable and for the most part do not pose any harm to humans or animals. Through the avoidance of high temperatures and pressures and of large consumptions of metals and organic solvents, the generation of waste per unit of product is drastically reduced.

Advanced Pharmaceutical Intermediates (APIs)

The most important property of a catalyst for application in a process is not its activity but rather its selectivity, followed by its stability, which is just activity integrated over time. In comparison with other catalysts, enzymes often feature superior selectivity, especially regio- and enantioselectivity. Enzymes are destined for selective synthesis of molecules with several similar functional groups or chiral centers. A growing emphasis is laid on the synthesis of enantiomerically pure compounds (EPCs). The interest in EPCs shown by all areas of the life science industries (e.g., pharmaceuticals, food and agriculture), stems from the challenge to develop structurally optimised inhibitors, almost always containing chiral centers.

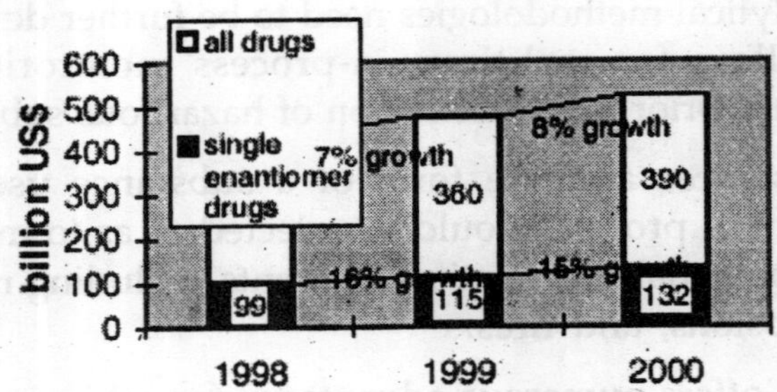

Figure 3. Relevance of chiral drugs.

The importance of the drive for enantiomeric purity for biocatalysis can be discerned from the fact that fully 88% of all biocatalytic processes above a scale of 100 kg involve chiral targets. Of the new drugs introduced in 2000, 76% were single enantiomers compared with 21% in 1991. As Figure 3 reveals, the revenues from single-enantiomer drugs have exceeded $ (or •) 100 billion in 2000, compared to $ (•) 500 billion total, and feature a higher growth rate than non-single-enantiomer drugs (155 vs. 8% annual growth).

Penetration of Biocatalysis

Biotechnology is an old science, or even an old art. The big ideas and driving forces for biocatalysis in the 20th century were twofold: first, the idea of catalysis as transition-state complementarity in 1944 and second, the development of molecular biology after 1978 to allow the design of enzymes and their production vehicles.

Even a few years ago, biocatalytic processes on an industrial scale were few. As with so many novel technologies, however, the time lag between research, development, and large-scale application just had to pass before we could witness a range of such processes in industrial practice today. In a review summarising the status of biocatalytic processing in industry, Straathof records that the number of industrial-scale biocatalytic processes has more than doubled over the last 10 years. Straathof included 134 industrial-scale biotransformations, counting processes on a scale of > 100 kg with whole cells or enzymes starting from a precursor other than a C-source.

3

Expression of Biopharmaceuticals

Biopharmaceuticals are large molecules produced by living cells, which are currently the mainstay products of the biotechnology industry. Indeed, biologics such as Genentech's human growth factor somatropin or Amgen's recombinant erythropoietin (EPO) have shown that biopharmaceuticals can benefit a huge number of patients and also generate big profits for these companies at the same time. The single most lucrative product is EPO and combined sales of the recombinant EPO products "Procrit" (Ortho biotech) and "Epogen" (Amgen) have reportedly surpassed the $6.5 billion mark.

But it has also become obvious over the last couple of years that current fermentation capacities will not be sufficient to manufacture all biopharmaceuticals (in the market already or in development), because the market and demand for biologics is continuously and very rapidly growing; for antibodies alone (with at least 10 monoclonal antibodies approved and being marketed), the revenues are predicted to expand to US$3 billion in 2002 and US$8 billion in 2008. The 10 monoclonal antibodies on the market consume more than 75% of the industry's manufacturing capability. And there are up to 60 more that are expected to reach the market in the next six or seven years.

Altogether, there are about 1200 protein-based products in the pipeline with a 20% growth rate and the market for current and late stage (Phase III) is estimated to be US$42 billion in 2005 and even US$100 billion in 2010. But, there are obvious limitations of large-scale manufacturing resources and production capacities and pharmaceutical companies are competing. To circumvent this capacity crunch, it is necessary to look into other technologies rather than the established ones, like, for example, Escherichia coli or CHO (Chinese hamster ovary) cell expression. One solution to avoid these limitations could be the use of transgenic plants to express recombinant proteins at low cost, in GMP (good manufacturing practice) quality greenhouses (with purification and fill finish in conventional facilities).

Plants therefore provide an economically sound source of recombinant proteins, such as industrial enzymes, and biopharmaceuticals. Furthermore, using the existing infrastructure for crop cultivation, processing, and storage will reduce the amount of capital investment required for commercial production. For example, it was estimated that the production costs of recombinant proteins in plants could be between 10 and 50 times lower than those for producing the same protein in E. coli and Alan Dove describes a factor of thousand for cost of protein (US dollar per gram of raw material) expressed in, for example, CHO cells compared to transgenic plants. So, at the dawn of this new millennium, a solution is imminent to circumvent expression capacity crunches and to supply mankind with the medicines we need. Providing the right amounts of biopharmaceuticals can now be achieved by applying our knowledge of modern life sciences to systems that were on this planet long time before us - plants.

Currently, CHO cells are the most widely used technology in biomanufacturing because they are capable of expressing eukaryotic proteins (processing, folding, and posttranslational modifications) that cannot be provided by E. coli. A long track record exists for CHO cells, but unfortunately they bring some problems along when it comes to scaling up production. Transport of oxygen (and other gases) and nutrients is critical for the fermentation process, as well as the fact that heat must diffuse evenly to all cultured cells.

According to the Michaelis-Menten equation, the growth rate depends on the oxygen/nutrient supply; therefore, good mixing and aeration are a prerequisite for the biomanufacturing process and are usually achieved by different fermentation modes (Fig. 1). But the laws of physics set strict limits on the size of bioreactors. For example, an agitator achieves good heat flow and aeration, but with increased fermenter size, shear forces also increase and disrupt the cells - and building parallel lines of bioreactors multiplies the costs linearly.

A 10 000-L bioreactor costs between US$ 250 000 to 500 000 and takes five years to build (conceptual planning, engineering, construction, validation, etc.). An error in estimating demand for, or inaccurately predicting the approval of, a new drug can be incredibly costly. To compound the problem, regulators in the United States and Europe demand that drugs have to be produced for the market in the same system used to produce them for the final round of clinical trials, in order to guarantee bioequivalence (e.g. toxicity, bioavailability, pharmacokinetics, and pharmacodynamics) of the molecule. So, companies have to choose between launching a product manufactured at a smaller development facility (and struggling to meet market demands) or building larger, dedicated facilities for a drug that might never be approved!

Therefore, alternative technologies are used for the expression of biopharmaceuticals, some of them also at lower costs involved. One such alternative is the creation of transgenic animals ("pharming"), but this suffers from the disadvantage that it requires a long time to establish such animals (approximately 2 years). In addition to that, some of the human biopharmaceuticals could be detrimental to the mammal's health, when expressed in the mammary glands. This is why ethical debates sometimes arise from the use of transgenic mammals for production of biopharmaceuticals.

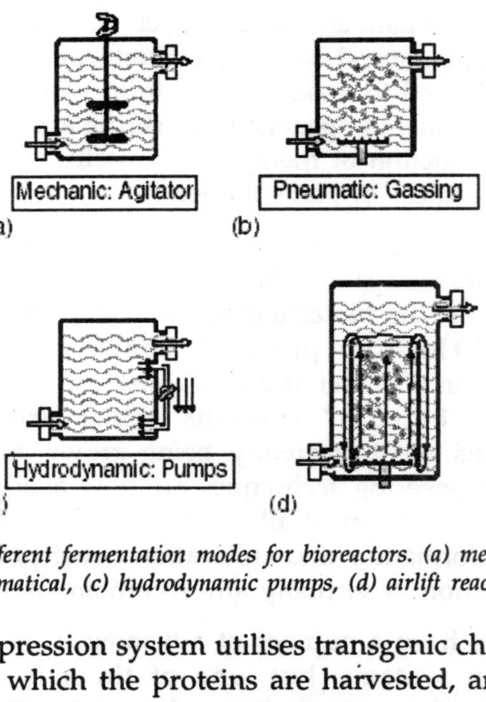

Figure 1. Different fermentation modes for bioreactors. (a) mechanical, (b) pneumatical, (c) hydrodynamic pumps, (d) airlift reactor.

Another expression system utilises transgenic chicken. The eggs, from which the proteins are harvested, are natural protein-production systems. Intensive animal housing constraints also make them more susceptible to disease.

In the light of development time, experience, costs, and ethical issues, plants are therefore the favored technology, since such systems usually have short gene-to-protein times (weeks), some are already well established, and as mentioned before, the involved costs are comparatively low. This low cost of goods sold (COGS) for plantderived proteins is mainly due to low capital costs: greenhouse costs are only US$ 10 per m^2 versus US$ 1000 per m^2 for mammalian cells.

Plants have been a source of medicinal products throughout human evolution. These active pharmaceutical compounds have been primarily small molecules, however. One of the most popular examples is aspirin (acetylsalicylic acid) to relieve pain and reduce fever. A French pharmacist first isolated natural salicin (a chemical relative of the compound used to make aspirin) from white willow bark in 1829. Advances in genetic engineering are now allowing for the production of therapeutic proteins (as opposed to small molecules) in plant tissues.

Expression of recombinant proteins in plants has been well documented since the 1970s and has slowly gained credibility in the biotechnology industry and regulatory agencies. The first proof of concept has been the incorporation of insect and pest resistance into grains. For example, "Bt corn" contains genes from Bacillus thuringensis and is currently being grown commercially. Genetic engineering techniques are now available for the manipulation of almost all commercially valuable plants. Easy transformation and cultivation make plants suitable for production of virtually any recombinant protein.

Plants have a number of advantages over microbial expression systems, but one of them is of outmost importance: they can produce eukaryotic proteins in their native form, as they are capable of carrying out posttranslational modifications required for the biological activity of many such proteins. These modifications can be acetylation, phosphorylation, and glycosylation, as well as others. Per se, there is no restriction to the kind of proteins that can be expressed in plants: vaccines (e.g. pertussis or tetanus toxins), serum proteins (e.g. albumin), growth factors (e.g. vascular endothelial growth factor (VEGF), erythropoietin), or enzymes (e.g. urokinase, glucose oxidase, or glucocerebrosidase).

However, enzymes sometimes have very complex cofactors, which are essential for their catalytic mode of action, but cannot be supplied bymost expression systems. This is why, for the expression of some enzymes, expression systems with special features and characteristics need to be developed. Another very important class of proteins is the antibodies (e.g. scFv, Fab, IgG, or IgA). More than 100 antibodies are currently used in clinical trials as therapeutics, drug delivery vehicles, in diagnostics and imaging, and in drug discovery research for both screening and validation of targets. Again, plants are considered as the system of choice for the production of antibodies ("plantibodies") in bulk amounts at low costs.

Since the initial demonstration that transgenic tobacco (Nicotiana tabacum) is able to produce functional IgG1 from mouse, full-length antibodies, hybrid antibodies, antibody fragments (Fab), and single-chain variable fragments (scFv) have been expressed in higher plants for a number of purposes. These antibodies can serve in health care and medicinal applications, either directly by using the plant as a food ingredient or as a pharmaceutical or diagnostic reagent after purification from the plantmaterial. In addition, antibodiesmay improve plant performance, for example, by controlling plant disease or by modifying regulatory and metabolic pathways.

Plant Expression Systems

When analysed the different expression systems regarding their strengths, weaknesses, opportunities, and threats

(SWOT), the advantages of plants and their potential to circumvent the worldwide capacity limitations for protein production became quite obvious (Fig. 2). Comparison of transgenic animals, mammalian cell culture, plant expression systems, yeast, and bacteria shows certain advantages for each of the systems. In the order in which the systems were just mentioned, we can compare them in terms of their development time (speed).

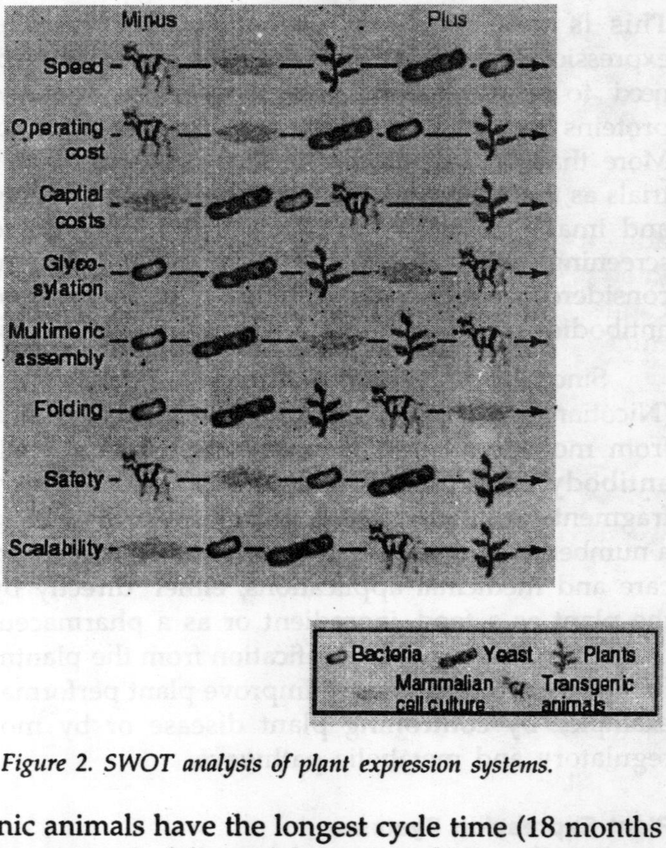

Figure 2. SWOT analysis of plant expression systems.

Transgenic animals have the longest cycle time (18 months to develop a goat), followed by mammalian cell culture, plants, yeast, and bacteria (one day to transform E. coli). If one looks at operating and capital costs, safety, and scalability, the data show that plants are beneficial: therefore, in the comparison (Fig. 2), they are shown on the right-hand side already. But even for glycosylation, multimeric assembly and folding (where plants are not shown on the right-hand side, meaning other systems are advantageous), some plant expression systems are moving in that direction.

An example of this is the moss system from the company greenovation Biotech GmbH (Freiburg, Germany). This system performs proper folding and assembly of even such complex proteins like the homodimeric VEGF. Even the sugar pattern could successfully be reengineered from plant to humanlike glycosylation. In addition to the potential of performing human glycosylation, plants also enjoy the distinct advantage of not harboring any pathogens, which are known to harm animal cells (as opposed to animal cell cultures and products); nor do the products contain any microbial toxins, TSE (Transmissible Spongiform Encephalopathies), prions, or oncogenic sequences.

In fact, humans are exposed to a large, constant dose of living plant viruses in the diet without any known effects/illnesses. Plant production of protein therapeutics also has advantages with regard to their scale and speed of production. Plants can be grown in ton quantities, be extracted with industrial-scale equipment, and produce kilogram-size yields from a single plot of cultivation. These economies of scale are expected to reduce the cost of production of pure pharmaceutical-grade therapeutics by more than 2 orders of magnitude versus current bacterial fermentation or cell culture reactor systems (plus raw material COGS are estimated to be as low as 10% of conventional cell culture expenses).

Although a growing list of heterologous proteins were successfully produced in a number of plant expression systems with their manifold advantages, there are also obvious downsides. One weakness is that no product has been approved for the market yet (but will be soon, since some are in Phase III clinical trials already, see Table 1). The other weakness is that no final regulatory guidelines exist. But as mentioned before, regulatory authorities (Food and Drug Administration (FDA), European Medicine Evaluation Agency (EMEA), and Biotechnology Regulatory Service (BRS) and the Biotechnology Industry Organisation (BIO) have drafted guidelines on plant-derived biopharmaceuticals (Table 2) and have asked the community for comments.

The FDA has also issued several PTC (Points To Consider) guidelines about plant-based biologics, and review of the July 2002 PTC confirms that the FDA supports this field and highlights the benefits of plant expression systems - including the absence of any pathogens to man from plant extracts. The main concerns of using plant expression systems are societal ones about environmental impacts, segregation risk, and contamination of the food chain.

But these threats can be dealt with, using nonedible plants (nonfood, nonfeed), applying advanced containment technologies (GMP greenhouses, bioreactors) and avoiding open-field production. Owing to the obvious strengths of plant expression systems, there has been explosive growth in the number of start-up companies. Since the 1990s, a number of promising plant expression systems have been developed, and in response to this "blooming field" big pharmaceutical companies have become more interested. Now, the plant expression field is "ripe" for strategic alliances, and, in fact, the last year has seen several major biotech companies begin partnerships with such plant companies.

The selection of several such partnerships shown in Table 1 clearly demonstrates that, in general, there has been sufficient experimentation with various crops to provide the overall proof of concept that transgenic plants can produce biopharmaceuticals. However, and this can be seen in the table as well, the commercial production of biopharmaceuticals in transgenic plants is still in the early stages of development and yet the most advanced products are in Phase III clinical development.

For a number of reasons, including the knowledge base developed on genetically modifying its genome, industrial processes for extracting fractionated products and the potential for large-scale production, the preferred plant expression system has been corn. However, the use of corn touches on a potential risk: some environmental activist groups and trade associations are concerned about the effect on the environment and possible contamination of the food supply. These issues are reflected in the regulatory guidelines and have been the driving force to investigate other plants as well.

While many mature and larger companies have been working in this area for many years, there are a number of newcomers that are developing expertise as well. These smaller companies are reacting to the concerns by looking at the use of nonedible plants that can be readily raised in greenhouses. All potential risks have to be assessed and contingency measures need to be established. Understanding the underlying issues is mandatory to make sophisticated decisions about the science and subsequently on the development of appropriate plant expression systems for production of biopharmaceuticals.

Ongoing public fears from the food industry and the public, particularly in Europe ("Franken Food") could have spillover effects on plant-derived pharmaceuticals. Mistakes and misunderstandings have already cost the genetically enhanced grain industry hundreds of millions of dollars. The only way to prevent plant expression systems from suffering the same dilemma is to provide the public with appropriate information on emerging discoveries and newly developed production systems for biopharmaceuticals. Real and theoretical risks involve the spread of engineered genes into wild plants, animals, and bacteria (horisontal transmission). For example, if herbicide resistance was transmitted to weeds, or antibiotic resistance was to be transmitted to bacteria, superpathogens could result.

If these genetic alterations were transmitted to their progeny (vertical transmission), an explosion of the pathogens could cause extensive harm. An example of this occurred several years ago, when it was feared that pest-resistant genes had been transmitted from Bt corn to milkweed - leading to the widespread death of Monarch butterflies. Although this was eventually not found to be the case, the public outcry over the incident was a wake-up call to the possible dangers of transgenic food technology. To avoid the same bad perception for biopharmaceuticals expressed in plants, there is the need for thorough risk assessment and contingency planning.

One method is the employment of all feasible safety strategies to prevent spreading of engineered DNA (genetic drift), like a basic containment in a greenhouse environment. Although no practical shelter can totally eradicate insect and rodent intrusion, this type of isolation is very effective for self-pollinators and those plants with small pollen dispersal patterns. The use of species-specific, fragile, or poorly transmissible viral vectors is another strategy. Tobacco mosaic virus (TMV), for example, usually only infects a tobacco host.

It requires an injury of the plant to gain entry and cause infection. Destruction of a field of TMV-transformed tobacco requires only plowing under or application of a herbicide. These factors prevent both horisontal and vertical transmission. In addition, there is no known incidence of plant viruses infecting animal or bacterial cells. Another approach is to avoid stable transgenic germlines and therefore most uses of transforming viruses do not involve the incorporation of genes into the plant cell nucleus. By definition, it is almost impossible for these genes to be transmitted vertically through pollen or seed.

The engineered protein product is produced only by the infected generation of plants. Another effective way to reduce the risk of genetic drift is the use of plants that do not reproduce without human aid. The modern corn plant cannot reproduce without cultivation and the purposeful planting of its seeds. If a plant may sprout from grain, it still needs to survive the wintering-over process and gain access to the proper planting depth. This extinction process is so rapid, however, that the errant loss of an ear of corn is very unlikely to grow a new plant.

Another very well-known example of self-limited reproduction is the modern banana. It propagates almost exclusively through vegetative cloning (i.e. via cuttings). Pollination is the natural way for most plants to spread their genetic information, make up new plants, and to deliver their offspring in other locations. The use of plants with limited range of pollen dispersal and limited contact with compatible wild hosts therefore is also very effective to prevent genetic drift. Corn, for example, has pollen, which survives for only 10 to 30 min and, hence, has an effective fertilising radius of less than 500 m. In North America, it has no wild-type relatives with which it could cross-pollinate.

In addition to being spatially isolated from nearby cornfields, transgenic corn can be "temporally isolated" by being planted at least 21 days earlier or 21 days later than the surrounding corn, to ensure that the fields are not producing flowers at the same time. Under recent USDA

(U.S. Department of Agriculture) regulations, the field must also be planted with equipment dedicated to the genetically modified crop. For soybeans, the situation is different, since they are virtually 100% self-fertilisers and can be planted in very close proximity to other plants without fear of horisontal spread. Another option is the design of transgenic plants that have only sterile pollen or - more or less only applicable for greenhouses - completely prevent cross-pollination by covering the individual plants.

One public fear regards spreading antibiotic resistance from one (transgenic donor) plant to other wild type plants or bacteria in the environment. Although prokaryotic promoters for antibiotic resistance are sometimes used in the fabrication and selection of transgenic constructs, once a transgene has been stably incorporated into the plant genome, it is under the control of plant (eukaryotic) promoter elements. Hence, antibiotic-resistance genes are unable to pass from genetically altered plants into bacteria and remain functional. As stated earlier, another common fear is the creation of a "super bug."

The chance of creating a supervirulent virus or bacterium from genetic engineering is unlikely, because the construction of expression cassettes from viral or bacterial genomes involves the removal of the majority of genes responsible for the normal function of these organisms. Even if a resultant organism is somewhat functional, it cannot compete for long in nature with normal, wild-type bacteria of the same species. As one can see from the aforementioned safety strategies, considerable effort is put into the reduction of any potential risk from the transgenic plant for the environment.

In general, the scientific risk can be kept at a minimum, if common sense is applied - in accordance with Thomas Huxley that "Science is simply common sense at its best." For example, protein toxins (for vaccine production) should never be grown in food plants. Additionally, the following can be employed as a kind of risk management to prevent the inappropriate or unsafe use of genetically engineered plants: An easily recognised phenotypic characteristic can be coexpressed in an engineered product (e.g. tomatoes that contain a therapeutic protein can be selected to grow in a colorless variety of fruit).

— Protein expression can be induced only after harvesting or fruit ripening. For example, CropTech's (Blacksburg, VA, USA) inducible expression system in tobacco, MeGA-PharM, leads to very efficient induction upon leaf injury (harvest) and needs no chemical inducers. This system possesses a fast induction response and protein synthesis rate, and thus leads to high expression levels with no aged product in the field (no environmental damage accumulation).

— Potentially antigenic or immunomodulatory products can be induced to grow in, or not to grow in, a certain plant tissue (e.g. root, leaf/stem, seed, or pollen). In this way, for example, farmers can be protected from harmful airborne pollen or seed dusts.

— Although no absolute system can prevent vandalism or theft of the transgenic plants, a very effective, cheap solution has been used quietly for many years now in the United States. Plots of these modified plants are being grown with absolutely no indication that they are different from a routine crop. In the Midwest, for example, finding a transgenic corn plot among the millions of acres of concurrently growing grain is virtually impossible. The only question here is, if this approach really helps facilitating a fair and an open discussion with the public. Asking the same question for the EU is not relevant: owing to labeling requirements, this approach would not be feasible, as, in general, it is much more difficult to perform open-field studies with transgenic plants.

Moving Plants to Humanlike Glycosylation

Plant production of therapeutic proteins has many advantages over bacterial systems. One very important feature of plant cells is their capability of carrying out posttranslational modifications. Since they are eukaryotes (i.e. have a nucleus), plants produce proteins through an ER (endoplasmatic reticulum) pathway, adding sugar residues also to the protein a process called glycosylation. These carbohydrates help determine the threedimensional structures of proteins, which are inherently linked to their function and their efficacy as therapeutics. This glycosylation also affects protein bioavailability and breakdown of the biopharmaceutical; for example, proteins lacking terminal sialic acid residues on their sugar groups are often targeted by the immune system and are rapidly degraded.

The glycosylation process begins by targeting the protein to the ER. During translation of mRNA (messenger RNA) into protein, the ribosome is attached to the ER, and the nascent protein fed into the lumen of the ER as translation proceeds. Here, one set of glycosylation enzymes attaches carbohydrates to specific amino acids of the protein. Other glycosylation enzymes either delete or add more sugars to the core structures. This glycosylation process continues into the Golgi apparatus, which sorts the new proteins, and distributes them to their final destinations in the cell (Fig. 3).

Bacteria lack this ability and therefore cannot be used to synthesise proteins that require glycosylation for activity. Although plants have a somewhat different system of protein glycosylation from mammalian cells, the differences usually prove not to be a problem.

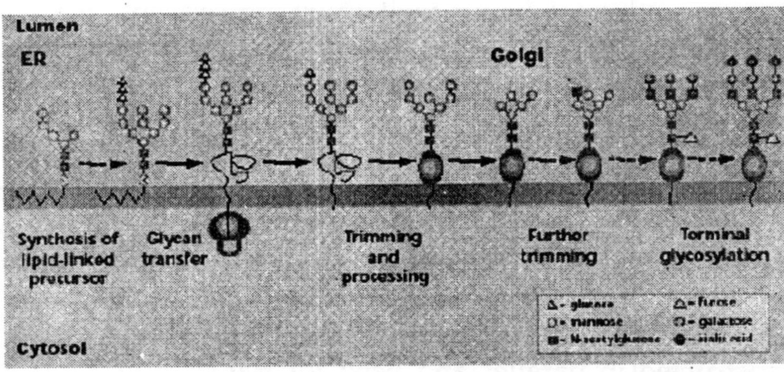

Figure 3. The glycosylation pathway via ER and Golgi apparatus.

Some proteins, however, require humanlike glycosylation - they must have specific sugar structures attached to the correct sites on the molecule to be maximally effective. Therefore, some efforts are being made in modifying host plants in such a way that they provide the protein with human glycosylation patterns. One example of modifying a plant expression system in this way is the transgenicmoss, which will be described below.

Tobacco and Moss Plants

All systems share the advantage of utilising nonedible plants (nonfood and nonfeed) and can be kept in either a greenhouse or a fermenter to avoid any segregation risk. Another obvious advantage is secretion of the protein into the medium so that no grinding or extraction is required. This is very important in light of downstream processing: protein purification is often as expensive as the biomanufacturing and should never be underestimated in the total COGS equation.

Phytomedics (Dayton, NJ, USA) uses tobacco plants as an expression system for biopharmaceuticals. Besides the advantage of being well characterised and used in agriculture for some time, tobacco has a stable genetic system, provides highdensity tissue (high protein production), needs only simple medium, and can be kept in a greenhouse. Optimised antibody expression can be rapidly verified using transient expression assays (short development time) in the plants before creation of transgenic suspension cells or stable plant lines (longer development time). Different vector systems, harboring targeting signals for subcellular compartments, are constructed in parallel and used for transient expression.

Applying this screening approach, high expressing cell lines can rapidly be identified. For example, transgenic tobacco plants, transformed with an expression cassette containing the GFP (Green Fluorescent Protein) gene fused to an aps (amplification promoting sequence), had greater levels of corresponding mRNAs and expressed proteins compared to transformants lacking aps. Usually, downstream processing (isolation/ extraction and purification of the target protein) is limiting for such a system, for example, if the protein has to be isolated from biochemically complex plant tissues (e.g. leaves), this can be a laborious and expensive process and a major obstacle to large-scale protein manufacturing.

To overcome this problem, secretion-based systems utilising transgenic plant cells or plant organs aseptically cultivated in vitro would be one solution. However, in vitro systems can be expensive, slow growing, unstable, and relatively low yielding. This is why another interesting route was followed. Secretion of molecules is a basic function of plant cells and organs in plants, and is especially developed in plant roots. In order to take up nutrients from the soil, interact with other soil organisms, and defend themselves against numerous pathogens, plant roots have evolved sophisticated mechanisms based on the secretion of different biochemicals (including proteins like toxins) into their neighborhood (rhizosphere).

In fact, Borisjuk and coworkers could demonstrate that root secretion can be successfully exploited for the continuous production of recombinant proteins in a process termed "rhizosecretion." Here, an endoplasmic reticulum signal peptide is fused to the recombinant protein, which is then continuously secreted from the roots into a simple hydroponic medium (based on the natural secretion from roots of the intact plants). The roots of the tobacco plant are sitting in a hydroponic tank, taking up water and nutrients and continuously releasing the biopharmaceutical. By this elegant set up, downstream processing becomes easy and cost-effective, and also offers the advantage of continuous protein production that integrates the biosynthetic potential of a plant over its lifetime and might lead to higher protein yields than single-harvest and extraction methods.

Protein-production System

ICON Genetics has developed a protein-production system that relies on rapid multiplication of viral vectors in an infected tobacco plant. Viral transfection systems offer a number of advantages, such as very rapid (1 to 2 week) expression time, possibility of generating initial milligram quantities within weeks, high expression levels, and so on. However, the existing viral vectors, such as TMV-based vectors used by, for example, Large Scale Biology Corp. (Vacaville, CA, USA) for production of single-chain antibodies for treatment of non-Hodgkin lymphoma (currently in Phase III clinical trials, Table 1), had numerous shortcomings, such as inability to express genes larger than 1 kb, inability to coexpress two or more proteins (a prerequisite for production of monoclonal antibodies, because they consist of the light and heavy chains, which are expressed independently and are subsequently assembled), low expression level in systemically infected leaves, and so on.

ICON has solved many of these problems by designing a process that starts with an assembly of one or more viral vectors inside a plant after treating the leaves with agrobacteria, which deliver the necessary viral vector components. ICON's proviral vectors provide advantages of fast and high-yield amplification processes in a plant cell, simple and inexpensive assembly of expression cassettes in planta, and full control of the process. The robustness of highly standardised protocols allows the use of inherently the same safe protocols for both laboratory-scale as well as industrial production processes. In this system, the plant is modified transiently rather than genetically and reaches the speed and yield of microbial systems while enjoying posttranslational capabilities of plant cells.

De- and reconstructing of the virus adds some safety features and also increases efficiency. There is no "physiology conflict," because the "growth phase" is separated from the "production phase," so that no competition occurs for nutrients and other components required for growth and also for expression of the biopharmaceutical at the same time. This transfection-based platform allows the production of proteins in a plant host at a cost of US$1 to 10 per gram of crude protein. The platform is essentially free from limitations (gene insert size limit, inability to express more than one gene) of current viral vector-based platforms.

The expression levels reach 5 g per kilogram of fresh leaf tissue (or some 50% of total cellular protein!) in 5 to 14 days after inoculation. Since the virus process (in addition to superhigh production of its own proteins, including the protein of interest) leads to the shutoff of the other cellular protein synthesis, the amount of protein of interest in the initial extract is extremely high. It thus results in reduced costs of downstream processing. Milligram quantities can be produced within two weeks, gram quantities in 4 to 6 months, and the production system is inherently scalable. Anumber of high-value proteins have been successfully expressed, including antibodies, antigens, interferons, hormones, and enzymes.

Greenovation Biotech GmbH has established an innovative production system for human proteins. The system produces pharmacologically active proteins in a bioreactor, utilising a moss (Physcomitrella patens) cell culture system with unique properties. It was stated before that posttranslational modifications for some proteins are crucial to gain complete pharmacological activity. Since moss is the only known plant system that shows a high frequency of homologous recombination, this is a highly attractive tool for production strain design. By establishing stable integration of foreign genes (gene knockout and new transgene insertion) into the plant genome, it can be programmed to produce proteins with modified glycosylation patterns that are identical to animal cells.

The moss is photoautotrophic and therefore only requires simple media for growth, which consist essentially of water and minerals. This reduces costs and also accounts for significantly lower infectious and contamination risks, but in addition to this, the system has some more advantages:

— The transient system allows production of quantities for a feasibility study within weeks - production of a stable expression strain takes 4 to 6 months.

— On the basis of transient expression data, the yield of stable production lines is expected to reach 30 mg L

1 per day. This corresponds to the yield of a typical fed-batch culture over 20 days of 600 mg L^{-1}. Bacterial fermentation usually requires addition of antibiotics. Formoss cultivation, no antibiotics are needed - this avoids the risk of traces of antibiotics having a significant allergenic potential in the finished product.

— Genetic stability is provided by the fact that the moss is grown in small plant fragments and not as protoplasts or tissue cultures avoiding somaclonal variation.

— As a contained system, the moss bioreactor can be standardised and validated according to GMP standards mandatory in the pharmaceutical industry.

— Excretion into the simple medium is anothermajor feature of the moss bioreactor, which greatly facilitates downstream processing.

The first step to get humanlike glycosylation in plants is to eliminate the plant glycosylation, for example, the attachment of β-1-2-linked xylosyl and α-1-3-linked fucosyl sugars to the protein, because these two residues have allergenic potential. Greenovation was able to knockout the relevant glycosylation enzymes xylosyl transferase and fucosyl transferase, which was confirmed by RT-PCR (reverse transcriptase PCR). And indeed, xylosyl and fucosyl residues were completely removed from the glycosylation pattern of the expressed protein as confirmed by MALDI-TOF (matrix assisted laser desorption ionization time of flight) mass spectroscopy analysis.

A very challenging protein to express is VEGF because this homodimer consists of two identical monomers linked via a disulfide bond. To produce VEGF in an active form, the following need to be provided:

— Monomers need to be expressed to the right level.

— Monomers need to be correctly folded.

— Homodimer needs to be correctly assembled and linked via a disulfide bond.

— Complex protein needs to be secreted in its active form.

The moss system adds no plant-specific sugars to the protein - a major step toward humanlike glycosylation. Furthermore, moss is a robust expression system leading to high yields at 15 to 25.C and the pH can be adjusted from 4 to 8 depending on the optimum for the protein of interest. Adapting existing technology for largescale cultivation of algae, fermentation of moss in suspension culture allows scaling of the photobioreactors up to several 1000 L (Fig. 11). Finally, the medium is inexpensive, since onlywater and minerals are sufficient.

Other Systems Used for Plant Expression

Several different plants have been used for the expression of proteins in plants. All these systems have certain advantages regarding edibility, growth rate, scalability; gene-to-protein time, yield, downstream processing, ease of use, and so on. A selection of different expression systems is listed:

Alfalfa	Ethiopian mustard	Potatoes
Arabidopsis	Lemna	Rice
Banana	Maize	Soybean
Cauliflower	Moss	Tomatoes
Corn	Oilseeds	Wheat

Some of these systems have been used for research on the basis of their ease of transformation, well-known characterisation, and ease to work with. However, they are not necessarily appropriate for commercial production.

Which crop is ultimately used for full-scale commercial production will depend on a number of factors including

— time to develop an appropriate system (gene-to-protein);

— section of the plant expressing the product/possible secretion;

— cost and potential waste products from extraction;

— "aged" product/ease of storage;

— long-term stability of the storage tissue;

— quantities of protein needed (scale of production).

Depending on the genetic complexity and ease of manipulation, the development time to produce an appropriate transgenic plant for milligram production of the desired protein can vary from 10 to 12 months in corn as compared to only weeks in moss. Estimates for full GMP production in corn are 30 to 36 months and approximately 12 months for moss. Expression of the protein in various tissues of the plant can result in a great variation in yield. Expression in the seed can often lead to higher yields than in the leafy portion of the plant. This is another explanation for the high interest in using corn, which has a relatively high seed-to-leaf ratio.

Extraction from leaf can be costly as it contains a high percentage of water, which could result in unavoidable proteolysis during the process. Proteins stored in seeds can be desiccated and remain intact for long periods of time. The purification and extraction of the protein is likely to be done by adaptations of current processes for the extraction and/or fractionation. For these reasons, it is anticipated that large-scale commercial production of recombinant proteins will involve grain and oilseed crops such as maize, rice, wheat, and soybeans.

On the basis of permits for open-air test plots issued by the USDA for pharmaceutical proteins and industrial biochemicals, corn is the crop of choice for production with 73% of the permits issued. The other major crops are soybeans (12%), tobacco (10%), and rice (5%). In general, the use of smaller plants that can be grown in greenhouses is an effective way of producing the biopharmaceuticals and alleviating concerns from environmental activist groups that the transgenic plant might be harmful to the environment (food chain, segregation risk, genetic drift, etc.).

Analytical Characterisation

Validated bioanalytical assays are essential and have to be developed to characterise the biopharmaceuticals during the production process (e.g. in-process control) and to release the final product for use as a drug in humans. These assays are applied to determine characteristics such as purity/ impurities, identity, quantity, stability, specificity, and potency of the recombinant protein during drug development. Since the very diverse functions of different proteins heavily depend on their structure, one very valuable parameter in protein characterisation is the elucidation of their three-dimensional structure.

Although over the last couple of years a lot of effort was put into a method for improving the elucidation of protein structures, it is still very time consuming to solve the 3-D structure of larger proteins. This is why despite the high degree of information that can be obtained from the protein structure, this approach cannot be applied on a routine basis. Therefore, tremendous efforts are put into the development of other assays to guarantee that a potent biopharmaceutical drug is indeed ready for use in humans.

4

Enzymes for Efficient Biocatalysis

Biocatalysts are being used extensively in the industrial production of bulk chemicals, pharmaceutical and agrochemical intermediates, active pharmaceuticals and food ingredients. Over 300 processes to date have been implemented in industry, demand for complex chiral drugs is high, and environmentally clean processes are increasingly required. Once it was realized that enzymes can also act within organic solvents containing little or no water, biocatalysis received a further boost. In fact, most synthetic uses of enzymes to date have involved the use of hydrolases in such systems.

In spite of the impressive number of applications (Table 1), enzymes do not exhibit satisfying performance or characteristics in terms of activity, stability and above all selectivity. Researchers have overcome some of these shortcomings by altering substrates (substrate engineering), by modifying reaction system (medium engineering) or by enzyme engineering. Enzyme engineering is based either on physical modifications of the protein or on modifications at the genetic (DNA) level. Microbial screening is a simple and frequently used method for finding new biocatalysts with required properties. These days it utilizes high-throughput-screening and metagenome analysis of uncultured microorganisms, with these and related approaches enabling researchers to take full advantage of the microbial diversity that exists.

Table 1. Properties and possibilities of enzyme applications

Enzyme	Main industrial application
Lipases	Interesterification of triacylglycerols, chiral synthesis
Glutaminase	Flavor-enhancing enzyme
Glycosynthases	Oligosaccharide synthesis
Epoxide hydrolases	Resolution of aryl- and substituted alicyclic epoxides
	Enantioselective catalyst
Monooxygenases,	Baeyer-Villiger oxidation
oxidizing enzymes	Oxidation,
Amylases	Pharmaceutical and fine-chemical industries
Oxidoreductases	Oxidation/reduction reactions
Lipoxygenases	Biosynthesis of inflammatory mediators
Nitrilase family	Different reactions

Also, extremozymes isolated from extremophiles are very attractive for industrial applications. Recently, it has been shown that small changes in the active site of enzymes enable new possibilities for catalyzing novel reactions. Bornscheuer and Kazlauskas described these properties in an excellent review about catalytic promiscuity.

During the last two decades variations in solvent systems have been used to influence the performance of enzymatic transformations. Enzymes in nonaqueous solvents can be active provided that the essential water layer around them is not stripped off. Medium engineering in the context of biocatalysis in nonaqueous media involves the modification of the immediate vicinity of the biocatalyst. Non-polar solvents are superior to polar ones since the former provide a better microenvironment for the enzyme, however, the solvent effects cannot be generalized too far. There are various exceptions, of which lipases are a particular case.

For instance, porcine pancreatic lipase is active in anhydrous pyridine, suggesting that it can retain the bound water even in a water-miscible solvent. Despite the fact that the effect of an organic solvent or an additive cannot be reliably predicted, addition of such solvents or additives has formed a useful approach for improving enzyme properties. For the Candida rugosa lipase-catalyzed hydrolysis of various substituted phenoxypropionates, the addition of 30-70 % dimethyl sulfoxide (DMSO) has resulted in increased enantioselectivity, (E = 4 to >100). Similar effects were also observed for the same reaction by adding aqueous sodium dodecyl sulfate (SDS) (E = 3.8 to >100) or by chemically modifying the lipase.

Recently, Iding et al. described the chemo-enzymatic preparation of chiral Nprotected 3-aminopyrrolidine derivatives at preparative scale, based on resolution of N-protected D,L-asparagine esters with proteases as a key step (Figure 1). The enzyme inhibition that occurred at high substrate levels in this process was overcome by the addition of co-solvents. In a classical example, Hirose et al. described the reversal of enantioselectivity with a change of solvent (Figure 2).

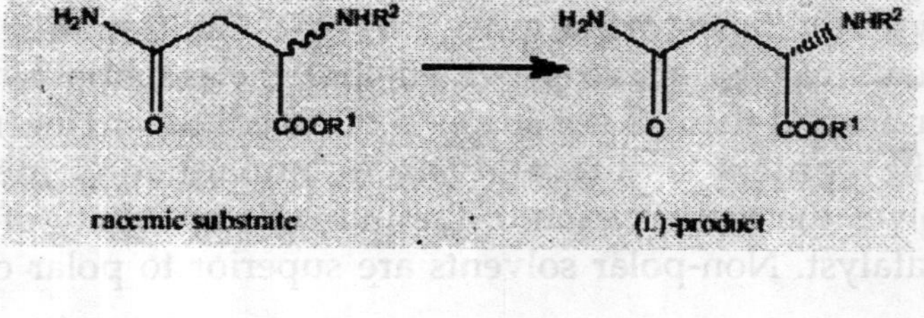

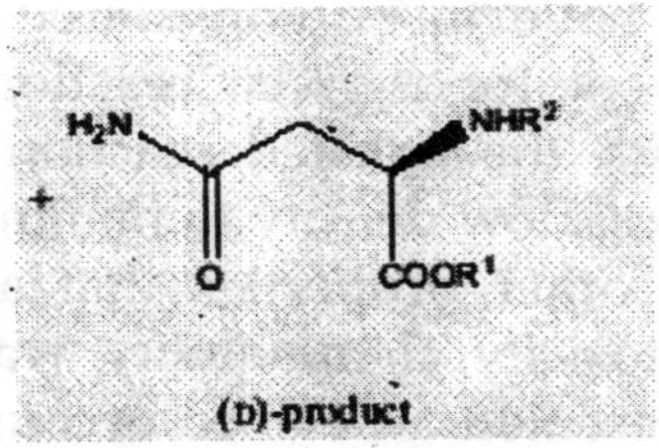

Figure 1. Resolution of N-protected D,L-asparagine esters

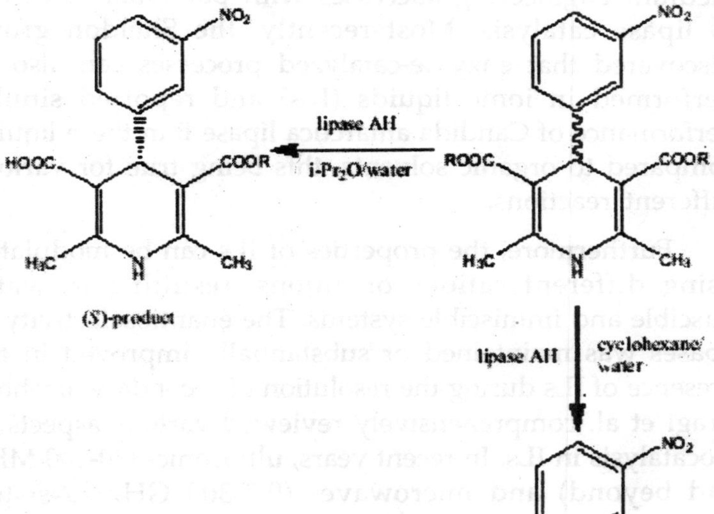

Figure 2. Complete solvent-dependent reversal of enantioselectivity in hydrolysis by Candida rugosa lipase in organic solvent saturated with water. Lipase AH: C. rugosa lipase; i-Pr$_2$O: diisopropyl ether

Several aspects have to be considered in choosing an appropriate solvent for a given reaction. These include compatibility with the selected reaction (substrates and products), inertness, low density to minimize mass transfer limitations, and other properties that are suitable (e.g. surface tension, toxicity, flammability, waste disposal and cost). Halling presented a detailed account of predictions that can be made to elucidate the influence of solvent selection on the equilibrium. Reetz reviewed various medium engineering successes with particular reference to lipase catalysis. Most recently, the Sheldon group discovered that enzyme-catalyzed processes can also be performed in ionic liquids (ILs) and reported similar performance of Candida antarctica lipase B in these liquids compared to organic solvents, this being true for various different reactions.

Furthermore, the properties of ILs can be modulated using different cations or anions, resulting in water miscible and immiscible systems. The enantioselectivity of lipases was maintained or substantially improved in the presence of ILs during the resolution of secondary alcohols. Kragl et al. comprehensively reviewed various aspects of biocatalysis in ILs. In recent years, ultrasonic- (20-100 MHz and beyond) and microwave- (0.3-300 GHz) assisted enzymatic reactions have been added to the repertoire of enzymology. A lot of attention has also been paid to biocatalysis in supercritical fluids or at high pressure.

Use of modified substrates has been investigated to improve efficiency of enzymatic reactions. Easy separation of the substrate and the product of a lipase-catalyzed acylation was achieved by using fluorinated acyl donors (Figure 3) in combination with an extraction step with perfluorinated solvents. This process has the advantages of easy scalability, a slightly higher enantioselectivity compared to the reaction performed with vinyl acetate as an acyl donor and a simple flash chromatographic separation.

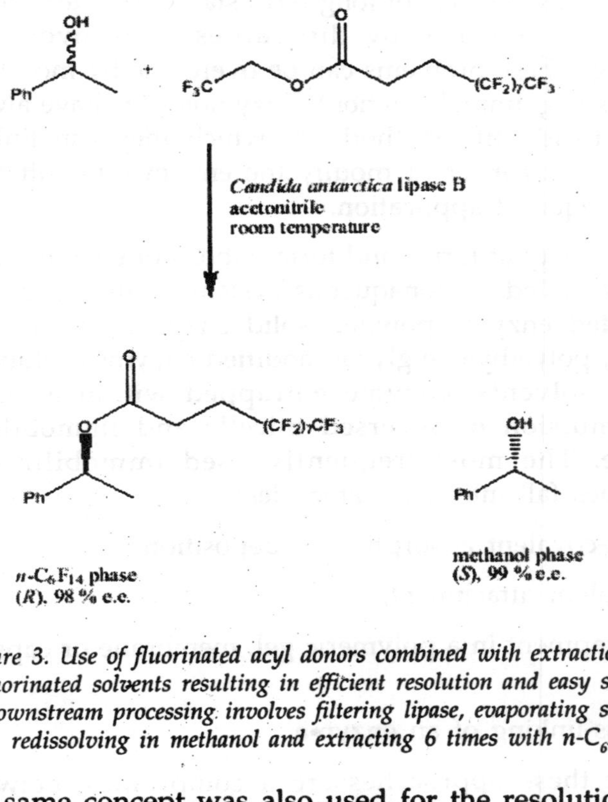

Figure 3. Use of fluorinated acyl donors combined with extraction using perfluorinated solvents resulting in efficient resolution and easy separation. Downstream processing involves filtering lipase, evaporating solvent, redissolving in methanol and extracting 6 times with n-C_6F_{14}.

The same concept was also used for the resolution of 1-(2-naphthyl)ethanol in a triphasic separative reaction. Creating variations in the substrate structure has also been tried in order to achieve improvements in biocatalysis. Conversion rates and yields, enantioselectivity, and biocatalyst stability were enhanced by using vinyl esters of chiral carboxylic acids instead of commonly used ethyl esters. The use of various substrate-engineering strategies in biohydroxylation has recently been reviewed.

Enzymes are versatile biocatalysts and find increasing application in many areas, including organic syntheses.

However, even after an enzyme has been identified as being useful for a certain reaction, its application is often hampered by the lack of long-term stability under process conditions and also by difficulties in recovery and recycling. These problems can be overcome by modifying enzymes (e.g. immobilization). Enzymologists have always strived to identify methods by which they can link an enzyme to a carrier or modify the enzymes to suit them for the required application.

A variety of forms and formulations of enzymes have been employed in nonaqueous systems: native enzymes, suspended enzyme powder, solid enzyme adsorbed on support, polyethylene glycol-modified enzymes soluble in organic solvents, enzyme entrapped within a gel or microemulsion or reversed micelle and immobilized enzyme. The most frequently used immobilization techniques fall into four categories:

(i) non-covalent adsorption or deposition,

(ii) covalent attachment,

(iii) entrapment in a polymeric gel, membrane or capsule, and

(iv) cross-linking of an enzyme.

Most of these approaches are a compromise between retaining high catalytic activity while achieving the advantages of immobilization. These techniques have resulted in biocatalysts exhibiting significantly higher stability than the native enzyme. Recent trends are focused on the use of new reagents/carriers and approaches that make use of the knowledge of enzyme structure and mechanism. These aspects have been explicitly dealt with very recently in an elegant review. However, a general and broadly applicable method for enzyme immobilization is still elusive and needs to be discovered. There are also no general guidelines for choosing the most appropriate form of the enzyme for a specific purpose.

Most often, the major criteria for enzyme immobilization are the price of the final immobilized enzyme and the simplicity of the immobilization procedure. Recently, Novozymes A/S presented a simple and cost-efficient procedure for Thermomyces lanuginosus lipase immobilization using silica granulation technology. This cheap immobilized lipase can be used for industrial interesterification reactions. Various developments have taken place in this area, including the generation of active and stable homogeneous (soluble) biocatalysts by covalent or noncovalent modification of the native enzyme. Covalent techniques are well described in the literature (e.g. the attachment of polyethylene glycol chains to enzymes).

Noncovalent modifications tend to be less common, although they are able to provide highly active and soluble enzyme forms. These developments have been well documented in various reviews. Enzymes are usually prepared from an aqueous/buffer solution via lyophilization and this results in undesirable changes in the protein's secondary structure, which could lead to denaturation of about 40 % of the active sites. The addition of specific small molecules (excipients) in the freeze-drying stage often improves catalytic activity. This formed the basis of molecular imprinting that involves formation of a complex between a macromolecule and a low-molecular-weight ligand in solution, followed by drying and washing with a selective solvent that removes the ligand.

The protein retains the ligand-induced conformation even after the removal of the ligand. Enzyme activation was achieved by adding crown ethers, lyoprotectants, transition-state analogues and substrate or substrate mimics or competitive inhibitors. Non-ligand lyoprotectants, such as sorbitol, sugars and PEG, also enhanced enzyme activity in organic solvents when added during lyophilization.

On the other hand, the addition of excipients to suspensions of native enzymes in organic solvents had no appreciable effect, indicating that the interaction of the excipient with soluble enzyme is essential to alleviate enzyme denaturation during lyophilization. Addition of surfactants or hydrophobic solgel materials before lyophilization enhanced the lipase activity in organic solvents by up to 100-fold. More dramatic was the activation of enzymes caused by adding an inorganic salt. Ion pairing of biocatalysts in the presence of very low concentrations of ionic surfactants resulted in remarkably active ion- -paired enzymes. Subtilisin and α-chymotrypsin with >1000-fold higher activities over native enzymes were generated.

Activated biocatalysts have already found application in the pharmaceutical industry. Salt-activated thermolysin (a bacterial protease) is used to acrylate taxol selectively in t-amyl alcohol. Commercially, Roche (Mannheim, Germany) and Novozymes A/S (Bagsvaerd, Denmark) market several adsorption-immobilized enzymes. Altus Biologics (Cambridge, MA, USA) offers various cross-linked enzyme crystals (CLECs) of different enzymes. Fluka (Buchs, Switzerland) supplies sol-gel entrapped enzymes.

With the onset of genome sequencing projects, bioinformatics is widely used to identify genes from different organisms and genome-wide comparisons form a major research area. With the modern tools of screening, high throughput screens, and availability of protein sequence databases and bioinformatics, screening for novel biocatalysts is pursued with increased vigor. Mathematical and data-mining tools have now been applied to analyze sequence-activity relationships of peptides and proteins and to assist in the design of proteins and peptides with specified properties.

Decreasing costs of DNA sequencing and methods to synthesize statistically representative sets of proteins allow modern heuristic statistics to be applied to protein engineering. This also provides an alternative approach to expensive assays or unreliable high-throughput surrogate screens. Biocatalysts and drugs from non-cultured microorganisms can also be developed from genomic information. The fact that only a fraction of the microbial flora in the environment can be cultured by standard techniques has led to the development of metagenomics, which is the genomic analysis of uncultured microorganisms. A function-driven approach or a sequence-driven approach is undertaken for this purpose and the results of such studies are already apparent from the discoveries of new antibiotics and novel enzymes.

Protein Engineering

Protein engineering has become an increasingly important strategy for improving enzymes despite intense competition from chemical modifications and related techniques. The application of protein engineering continues to grow at a very significant rate. Current strategies of protein engineering include rational design and directed evolution.

Rational Protein Design

Rational protein design was the earliest approach to engineering enzymes and is still widely employed to introduce the desired characteristics into a target protein. Crystallographers have already solved many enzyme structures and have paved the way to use computer modeling. The growing knowledge base on how to engineer certain basic enzyme properties (e.g. stability, activity and surface properties) is beginning to make rational design more efficient. Advances in rational design depend on the progress made in structure determination, improved modeling protocols and significant new insights into structure-function relationships.

Advances in modeling of free energy perturbation methods and molecular dynamics calculations also influence this area. There are numerous impressive cases of successful rational design. The literature on the rational engineering of biocatalyst stability has recently been reviewed. The enantioselectivity of the Candida antarctica lipase-catalyzed resolution of 1-chloro-2-octanol was improved (E = 14 to 28) by a single amino acid exchange (Ser-47 → Ala), as predicted by molecular modeling.

In another example, a transition-state stabilizing threonine near the active site was replaced by valine through site-directed mutagenesis, resulting in activity loss and the lost activity was restored by using chiral substrate (2-hydroxy-propanoate) containing the missing functional -OH group, which resulted in improved enantioselectivity (E = 1.6 to 22). Creating tailored enzymes with opposite enantiopreference may also be possible with this approach. Although numerous research groups are using modeling to predict enantioselectivity, much of the available data is still empirical and is not easily interpreted at the molecular level. This is largely due to ambiguity in identifying the parameters responsible for enzyme-substrate interactions.

In addition, properties like activity, specificity and stability are controlled by multiple sites on proteins. Substantial progress is essential in establishing these features, which would aid application of rational design in a truly directed manner. The importance of this exciting area of research was aptly reflected in a special issue of

Biochimica et Biophysica Acta: Protein Science and Molecular Enzymology that was devoted to protein engineering, spanning the advances achieved with various enzymes. There is another striking example of protein engineering success in relation to the activity of hydrolases on tertiary alcohol ester substrates (an important class of compounds that have not been accepted by numerous enzymes).

Pleiss et al. pointed out that lipases and esterases differ in a certain motif, which is responsible for the stabilization of the tetrahedral intermediate during catalysis. The motif could be either GX or GGGX (in single letter amino acid code; G is glycine, X is any amino acid; in a few cases the GGAX motif is the equivalent of the GGGX motif, where A is alanine) and is located in the oxyanion-binding pocket. Until recently this motif has only been described in a phenomenological manner. Bornscheuer's group was the first to identify the role of this motif in the activity towards tertiary alcohols. Only two lipases (Candida rugosa and C. antarctica lipase A) show activity on tertiary alcohol substrates.

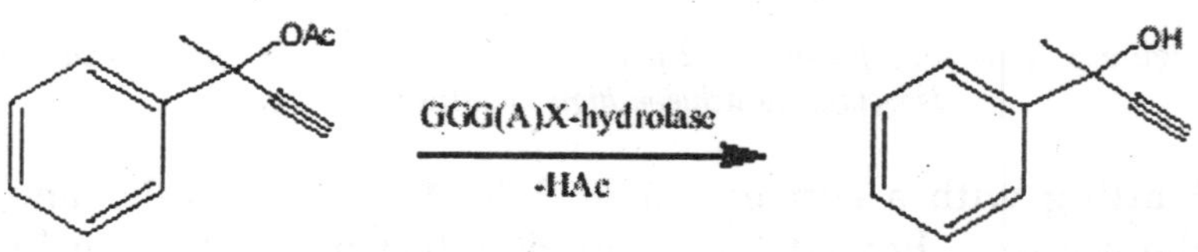

Figure 4. Activity of hydrolases towards tertiary alcohol substrates is determined by the amino acid motif: GGGX or GGAX.

Both C. rugosa and C. antarctica lipases, which contain the GGGX or the GGAX motif, and also numerous other GGGX-motif-containing hydrolases have been shown to be effective in hydrolyzing tertiary alcohol esters. Bacillus subtilis p-nitrobenzyl esterase (pNBE) was one of the GGGX motif-containing enzymes that showed moderate enantioselectivity. Through computer modeling, Gly-105 was predicted to be crucial for improving enantioselectivity and a mutant (Gly-105→ Ala) developed based on the modeling showed significant improvement (E = 6 to 19) in the resolution of (R,S)-2-phenyl-but-3-yn-2-yl acetate (Figure 4).

GXmotif hydrolases were all inactive towards the tertiary alcohols. In an extension of the study of this motif, Krishna et al. reported the dramatic resolution of the same substrate through transesterification. The reaction shown in Figure 4 also occurred to a certain extent by autohydrolysis, which circumvented high enantioselectivity. Therefore, the authors undertook transesterification of 2-phenyl-but-3-yn-2-ol (Figure 5) to avoid autohydrolysis and optimized reaction conditions for the same to achieve high enantioselectivity (E = 65).

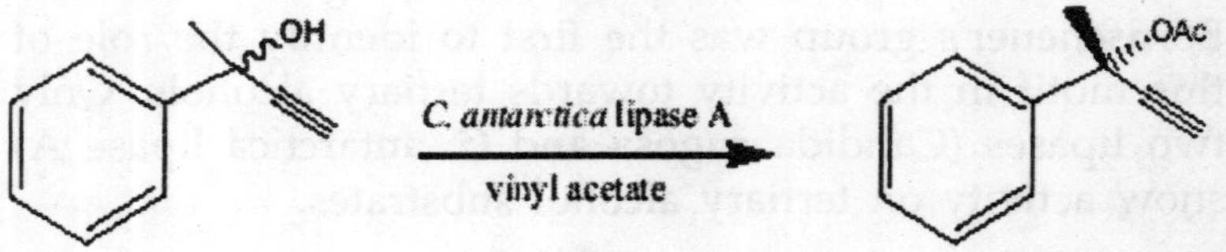

Figure 5. Transesterification of 2-phenyl-but-3-yn-2-ol with vinyl acetate in isooctane to achieve high enantioselectivity.

Starting with a conversion of 9 % (E = 22) in n-hexane, the authors showed the dramatic effect of the solvent by obtaining 21 % conversion (E = 49) in isooctane. By optimizing other parameters, the yield and enantioselectivity were further improved. While site-directed mutagenesis and rational protein design are widely practiced, an alternate method, directed evolution, is gaining increased attention from academic and industrial laboratories to modify and improve important biocatalysts; and to achieve various other objectives. This strategy combines random mutagenesis of the target gene with screening or selection for the desired property.

Thermostability is difficult to improve rationally and hence is a good target for directed evolution. Random mutagenesis and screening have resulted in more thermostable subtilisin and lipase. Contrary to these successes, a rational approach to increase the thermostability of Penicillium camembertii lipase by introducing a disulfide link was a failure. Tao and Cornish reviewed the milestones in directed evolution of enzymes. Numerous choices are available for creating DNA libraries, including error-prone PCR, combinatorial oli-gonucleotide mutagenesis, DNA shuffling, exon shuffling, random-priming recombination, random chimeragenesis on transient templates (RACHITT), staggered extension process (StEP recombination), heteroduplex recombination, incremental truncation for the creation of hybrids (ITCHY), recombined extension on truncated templates (RETT), degenerate oligonucleotide gene shuffling (DOGS) and in vivo recombination.

Lutz and Patrick discussed recently developed methods for directed evolution, including MAX randomization, which is an oligonucleotide-directed randomization method for creating molecular diversity in zinc-finger protein, and sequence saturation mutagenesis (SeSaM). The mutant genes obtained from any method of directed evolution are transformed and expressed in a suitable host. Enzyme libraries are selected and screened for a range of selection properties. Improved genes obtained in the first round may be used as templates for the subsequent evolution cycles.

For sexual evolution, a pool of homologous genes (or genes generated using asexual methods) is partially digested with DNase I and recombined by PCR. Alternatively, homologous recombination can also be achieved in vivo based on the transformation of Saccharomyces cerevisiae or Hansenula polymorpha. The linearized fragments possessing overlapping DNA sequences are cotransformed into the yeast. These fragments undergo homologous recombination in vivo, resulting in the restoration of the plasmid's circular topology. The plasmid repair is an efficient process and recombination between overlapping sequences as short as 30 bp can be readily achieved.

Several algorithms and computational models have been proposed to improve the efficiency of directed evolution. Improvements of enzymes by directed evolution include modifications of stability, activity and specificity. Bornscheuer subdivided examples for novel substrate specificity by directed evolution into three groups:

(i) conversion of substrates which in principle should be accepted by the enzyme under investigation, but do not react due to steric hindrance or electronic effects;
(ii) conversion of substrates which are usually converted by another enzyme;
(iii) conversion of compounds that are already converted by the enzyme of interest, but at very low activity compared to the well accepted substrates.

Synthesized oligonucleotides have been used to create suitably diverse gene libraries and to allow researchers to tailor the diversity of a library with higher precision. There are numerous success stories of directed evolution to list. DeSantis et al. developed a new technique called gene site saturation mutagenesis (GSSM), whereby each amino acid of the protein is replaced with each of the other 19 naturally occurring amino acids to create highly enantioselective nitrilase mutants. Although GSSM has led to the identification of novel mutants, it may not be a feasible technique for academic laboratories owing to the limited manpower and availability of instrumentation for each research area.

Many of the protein design efforts carried out to date have focused on using only a single method. While rational design is well suited to optimize direct interactions, it is not suitable for identifying distant mutations. Directed evolution, although good for optimizing an existing activity, may not be suitable for introducing a completely novel activity. Therefore, it is beneficial to combine the powers of individual techniques. Recently there have been attempts to combine rational design and directed evolution to improve biocatalysts. Also, directed evolution has been applied for metabolic pathway engineering.

The ability of directed evolution to transfer a large number of genes simultaneously has also formed the basis of the emerging area of combinatorial biocatalysis, which employs iterative reactions catalyzed by isolated enzymes or whole cells, in a natural or unnatural environment, on substrates in solution or on a solid phase and harnesses the natural diversity of enzymatic reactions to generate libraries of organic compounds. Very recently, researchers have had a breakthrough success in stringing together artificial DNA bases, which could aid the creation of altered genetic material for applications in medicine and technology. The tricky part is stitching these artificial bases into a chain, and getting the chains to replicate as natural DNA does, with the help of DNA polymerase, which works its way along a template strand of DNA, reading each base and collecting a matching base to create a new, copied strand.

5

Health Risk of Enzymes

Enzyme exposure mainly occurs from dust or liquid aerosols. The particles are deposited on the skin or on the mucous membranes of the respiratory tract. When an enzyme comes into contact with the respiratory tract or the skin, the body's immune system may be stimulated to produce antibodies resulting in respiratory allergy or contact urticaria, respectively. And because skin has a protein structure, enzymes which catalyse breakdown of proteins such as proteases are potential skin irritants.

Several common allergens from, e. g., house dust mites, storage mites, ragweed, pollens, Alternaria, cat dander, and bee venoms, are proteins with enzymatic activity. And if common antigens are enzymes by nature, classical natural (and artificial) enzymes may correspondingly represent antigens. Enzymes may also be more prone to cause sensitisation than other proteins because of their intrinsic capacity; e. g. the proteolytic function of many of these allergens has been proposed to be an important factor in the epithelial permeability and origin of allergy.

Sensitisation may follow minute exposure; even office personnel in enzyme-processing factories may become sensitised. High prevalence of sensitisation was also reported from the Danish enzyme industry, the baking industry, and from many other heavily exposed occupations. A discussion on the possibility and magnitude of medical problems caused by enzymes must consider the capacity of the individual enzyme to sensitise or irritate, its source and production method, its distribution, processing and stability, as well as time and extent of human exposure during production, industrial processing and consumption. Most medical reports come from the detergent industry.

The highest percentage of exposed people nowadays seems to be affected in bakeries, but sensitisation may also occur in the animal feed industry, resulting in concerns for meat consumers. And only recently several reports have been published on allergic reactions in other occupations (pharmaceutical industry, at laboratory work, etc.) and even in consumers after dermal, mucosal, oral and parenteral (injection, infusion) exposure to enzyme drugs, enzymes in cosmetics, soaps, in soft lens fluids, in meat, and other products of daily life. The critical importance of controlling exposure, and the consequences of not doing so effectively, are illustrated by early experience when enzymes were first used in laundry detergents.

Enzymes had been introduced in detergent products in 1963, and by 1969 80% of all laundry products contained enzymes. The detergent industry underestimated the risk of health effects in these early days. Early detergent enzymes were confectioned as powders and led to high airborne dust levels. As enzyme use increased, so did reports of respiratory illness at the workplace.

Enzymes play an important role both for occupational respiratory as well as for skin problems. Bakers, being exposed to flour containing 2 groups of allergens, the protein flour itself and natural enzymes, as well as enzymes added during the processing of the dough, have the highest risk of developing occupational contact urticaria in a Finnish study. In a German investigation, 23% of bakers suffering from workplace-related rhinitis and/or bronchial asthma were sensitised to industrial α-amylase products obtained from Aspergillus oryzae cultures.

Corresponding results were found in Switzerland, Italy, and Spain. Such reports again gave rise to public concern whether the addition of such products to food stuffs was safe for the consumer. Public awareness grew, encouraged by reports in both media and scientific literature. A dramatic rise of the percentage of allergic patients in the western world demands looking for specific reasons. The fact that classical allergens frequently are enzymes by their nature fuelled concerns regarding enzymes being dangerous products. And public discussion on the production of genetically modified plants and the introduction of enzymes produced by biotechnological methods boosted discussion on their safety for workers and consumers.

Allergy frequently also is an emotional problem, especially in people who feel sick without knowing the reason and suspect dangerous ingredients in the food to be the reason for this disease. The many reports on problems with enzymes seem to justify the fear, regarding that these biologically active substances exert their function in even minute amounts. And it is difficult to clear up the suspicion, if the proof of harmlessness is impossible to perform. In addition, continuing negative reports in the lay media being partially emotional but also substantiated by publications, that sensitisation still occurs despite encapsulation from the detergent industry and still increasing numbers of bakers, that have to leave their job because of work-related problems, keep the discussion going on. And indeed, the vast variety of different enzymes, being produced from many different sources and used not only in many fields of daily life, but also as drugs, makes an analysis of their potential health problems and how to deal with them essential.

Medical Problems

Reports on medical problems that are still linked to enzyme exposure (including additives and contaminants) are numerous. It is important to first define the diseases, their underlying mechanisms, and the different technical terms that are used in the medical literature. When an individual first encounters an antigen the cells of the immune system recognize this agent and either produce an immune reaction to it or become specifically tolerant depending on the circumstances. The immune reaction can take the form of cell-mediated immunity (delayed type or type IV reaction, the clinical correlate being allergic contact eczema) or may involve the production of antibodies directed towards the antigen (classically: immediate or type I hypersensitivity), the type of reaction being dependent on the way in which the antigen is presented to the lymphocytes.

Many immune reactions however display both kinds of response. On second and subsequent encounters with the antigen the type of response is largely determined by the outcome of the first antigenic challenge, but the quantity and the quality of the response are different. Whether a foreign substance leads to tolerance or sensitisation of any type (immediate or delayed) mostly depends on the type of agent; and in order to elucidate an immune response the agent must penetrate (or at least enter) the mucosal surface or the skin.

Enzymes modify skin antigens and thereby may reveal previously hidden or produce new antigens; but the clinical relevance of this finding is unknown. ZENTNER et al. postulated that the low molecular weight of proteolytic enzymes and their biologic activity might facilitate mucosal penetration more easily and thus compared to amylase and lipase permit an immune response and induction of allergic hypersensitivity. But whether a certain antigen in fact sensitises depends mostly on factors like molecular structure and binding capacities to certain reactive epitopes.

The knowledge on relevance and conclusions for practical use are currently limited. Initial studies could demonstrate, that in certain models using different animal strains antigenic activity varies quite considerably among different enzymes. Allergic diseases, that have been reported to be caused by enzymes in sensitised individuals, are numerous and may be assigned as follows: Contact with enzyme dust or aerosols may cause immediate type allergic reactions such as conjunctivitis (of the eyes), rhinitis and asthma. Upon direct skin contact either contact urticaria (antibody mediated immediate type reaction) or allergic contact dermatitis (delayed type reaction) may follow; both terms are frequently not well differentiated and many literature reports are not absolutely conclusive; transitional reactions or even indeterminable reactions, simulating classical immune reactions seem quite likely.

Tere are however conditions such as sensitisation that urgently need to be differentiated what is frequently not done in the literature nor in the clinical situation. Upon exposure, many people develop antibodies of different types: antigen-specific antibodies of the IgG-isotype do not correlate with disease activity but seem to be an indicator for exposure, i.e. detection of such antibodies does not signalise disease but just the amount of exposure (time and/or degree). In contrast, antigen-specific IgE-antibodies indicate allergic sensitisation, BUT NOT allergic disease:

this is an important difference, since in many clinical circumstances, where detection of IgE-antibodies is taken as an important parameter for allergic disease, this parameter is overvalued.

A person that is sensitised (has detectable IgE-antibodies in his/her blood; or a positive skin prick test) must not necessarily be sick (= be allergic). And whether this sensitisation will ever become relevant, i. e. the person will ever fall ill upon antigen exposure, is an open question that can only be answered statistically and depends on the specific antigen, the individuals constitution, the circumstances and other unknown and thus non-determinable reasons.

But without sensitisation no allergic disease, and thus sensitisation is a first step to clinical allergy although not a final one. Reports from the detergent, the baking and flour industry show that most exposed workers produce IgG-antibodies, several develop IgE-antibodies (become sensitised), but only a minority of them ever get sick and thus have to leave their workplace.

In contrast, in the production of car catalysts it is well known that workers sensitised to platinum should immediately leave their job as soon as antibodies are detected, since clinical disease (allergic asthma) can follow soon and may then persist even if further exposure is reduced. Schweigert et al. classified the clinical effects to detergent enzymes into three grades. Their experience showed, that asymptomatic, converted workers need not be removed from their work place. They were however placed under closer surveillance for the development of symptoms. Those with grade II or III symptoms had to be removed from the work area until their symptoms cleared and the source of exposure could be defined and eliminated.

Allergic Diseases

Respiratory allergy

Respiratory allergy is, regarding medical text books and clinical practice, divided into diseases of the upper respiratory tract (i. e. nose) and the lower respiratory tract (i. e. bronchi/lung). Allergic rhin (oconjunctiv) it is characterized by nasal itching, sneezing, watery rhinorrhoea, and nasal obstruction. Additional symptoms such as headache, impaired smell, and conjunctival symptoms may be associated. Patients with allergic rhinitis (hayfever) may suffer considerably as long as they are antigen-exposed (impaired quality of life parameters), and feel like suffering from a flu, but for weeks, months or even years.

Asthma bronchiale as defined by the World Health Organisation is a chronic condition characterised by recurrent bronchospasm, resulting from a tendency to develop reversible narrowing of the airways in response to various stimuli, and manifested by slowing of forced expiration which changes in severity, either spontaneously or as a result of therapy. This disease is not only associated with considerable morbidity, but also mortality. Despite of modern and effective treatment, or even because of this effective treatment, where many people rely too much on drug treatment and stop avoiding antigen exposure. In allergic asthma utmost priority should be given to antigen avoidance as complete as possible.

Skin allergy

Skin allergy is a frequently used term for contact allergy or allergic contact dermatitis. Regarding enzymes one should clearly differentiate two pathogenetically very different diseases, that may look quite similar, mimic each other in some situations, or even change their appearance from the one to the other form; i. e. classical allergic contact dermatitis (ACD) and immediate type contact urticaria (ACU) or protein contact dermatitis. Allergic contact dermatitis (ACD) is an inflammation of the skin, that occurs as a result of antigen contact with external agents.

ACD is normally the result of an immunological delayed-type reaction, clinically manifesting as eczema with considerable itching in the exposed areas following between 12 to 48 hours after skin contact with an allergen. The latter are usually low-molecular-weight chemicals that easily penetrate the skin and are presented by skin Langerhans cells to the immune system, resulting upon re-exposure in eczematous reactions. Typical antigens are used in various professional trades, such as dyes, household products or topically applied drugs and cosmetics.

Enzymes, being to large for penetrating normal skin, probably do not sensitise via this route and by these mechanisms, although there are a few case reports in the literature that seem to indicate this. But in all these studies, the way of sensitisation was not proven. Enzymes modify skin antigens, but whether this may contribute to skin sensitisation (delayed type reactions) is not proven. And indeed, the fact that those few reported persons with positive patch tests to non-proteolytic enzymes also have positive prick tests to the same enzymes may indicate that the observed patch-test reactions in fact are IgE-mediated allergic reactions and not necessarily the classical type-IV reactions. This may also indicate that these enzymes are causative agents in developing contact urticaria, and that this urticaria may proceed to an allergic contact dermatitis.

Allergic contact urticaria (ACU), in contrast to ACD, is the result of an immunological immediate type reaction (IgE-antibodies), that clinically manifests as immediate itching, swelling and urticarial lesions appearing immediately after antigen contact. In certain clinical situations, the picture might switch within days to an eczema, characterized by itching and mimicking ACD. This is typical in patients with atopic dermatitis (protein contact dermatitis), but also occurs in non-atopic individuals. The antigens are, similar to those in rhinitis and asthma, high-molecular weight antigens such as proteins, and enzymes are frequently described to cause ACU.

Contact urticaria is 'a new disease', that was first described only in the 1970s in fish mongers and vegetable merchants, that were handling such products over a prolonged period of time. In some case reports positive patch tests to non-proteolytic enzymes, i. e. cellulase, amylase, and xylanase were demonstrated. It is worth noting that these persons also have positive prick tests to the same enzymes. This may indicate that the observed patch-test reactions in fact are IgE-mediated allergic reactions and not necessarily the classical type-IV reactions, at least in some cases.

Since IgE-antibodies play an important role, people that easily produce IgE-antibodies so called atopics are especially prone to develop allergic diseases. Lesions might develop in contact areas only, but they may also generalize and cause rhino/conjunctivitis, asthma, urticaria, and even anaphylactic shock. Thus, IgE-mediated asthma can occur via skin contact or via inhalation.

Food allergy

Food allergy is certainly one of the most controversial and difficult areas among allergic diseases. A major reason for this situation is the lack of universal agreement on definitions and diagnostic criteria. Major clinical symptoms are urticaria and provocation of atopic dermatitis, bronchospasm, anaphylactic shock, vomiting, colics, diarrhoea, bloating, etc. For most scientists, food allergy encompasses all possible immunological reactions to foods, not only the classical IgE-mediated responses.

Drug allergy

Drug allergy is not a major issue of this summary although several enzymes used as drugs have been reported to cause allergic reactions, from discomfort, pruritus, to urticaria and angiooedema, serum sickness, bronchospasm, fever, arthralgias, myalgias, renal abnormalities to even fatal anaphylactic shock. Enzymes such as aprotinin, chymopapain, streptokinase, trypsin and others are used for the treatment of several medical problems; a report on IgE-mediated allergic reaction to hyaluronidase in paedriatic oncological patients may serve as example: Five out of 16 patients with tumors of the central nervous system (CNS) treated with hyaluronidase in addition to chemotherapeutic agents developed symptoms of immediate type allergic reaction to the hyaluronidase preparation from bovine testes.

Test Methods for Allergic Diseases

Test methods for allergic disease: Diagnostic procedures should always be based on a detailed case history and the clinical manifestation. Dependent on the pathomechanism of the reaction and the quality of antigen, different steps have to follow. Objective evaluation can be performed by skin tests and/or by serological allergen-specific IgE determination and/or by provocation tests.

For immediate type reactions such as hay fever, asthma and urticaria, skin prick testing with the suspected antigen (i. e. enzyme) in an appropriate concentration is the most direct approach to diagnosis. Skin testing, representing an in vivo-test method, is a fast, simple, cheap and straight forward test; but one has to be aware of the limitations (several reasons for false-negative or false-positive reactions), the need of control population when dealing with non-standardized products, and the danger of severe and even life-threatening reactions.

Skin prick tests, when being performed as titration test, may allow to follow the exposed person and early detect sensitisation or progression of disease activity skin prick tests are therefore frequently used to monitor enzyme-exposed persons. These skin prick tests, when performed with suitable reagents, are most sensitive but not entirely specific indicators for identifying symptomatic workers. For prick testing, commercial enzymes or non-commercial enzyme preparations can be used. For example, to reveal contact urticaria to α-amylases, two bacterial α-amylases and two fungal α-amylases were used in a Finnish study.

Dry enzyme preparations were extracted in 0.1 M potassium phosphate buffer, pH 7.4 and diluted to the Coca solution (0.5% sodium chloride, 0.3% sodium hydrogen carbonate, 0.4% phenol) to achieve a protein concentration of 200 µg/ml. A part of this solution (2.5 ml) was passed through a Millex-GV filter containing a 0.22-µm membrane into a sterile vial containing 2.5 ml glycerol, yielding a final protein concentration of 100 µg/ml. The Coca-glycerol solution served as a negative control. This description gives an example, how test preparations should be produced. In a routine work-up, preparations are frequently used as they are brought by the patient to the test institution, and tests are therefore frequently not only not standardized, but the results also non-reproducible.

In vitro-testing for antigen-specific antibodies of the IgE-isotype provides an important parameter. These tests are also quite specific but sometimes less sensitive than skin tests. In addition, commercial tests are hardly available (exceptions are papain and bromelain); self-prepared test reagents bear the likelihood of lacking reference values and standardization. Here again, reasons for false-negative (antibodies not detectable despite sensitisation or even clinical disease) and false-positive reactions (antibodies detectable, but not clinically relevant; i. e. sensitisation) are frequent. Bronchial provocation testing for asthma is not only time-consuming, but also tricky in performance and potentially dangerous in sensitised persons.

In addition, many asthmatics suffering from severe bronchial hypersensitivity react non-specifically positive if e. g. flour is inhaled due of the flours irritating activity on the bronchial mucosa. Regarding food allergy, even double-blind, placebo-controlled provocation procedures, often considered as 'gold standard' in diagnosis, is influenced by subjective evaluation and may require conditions not realised under experimental allergen procedures.

For delayed type reactions, patch testing on non-diseased skin is a reliable procedure. Since no commercial test substances are available, dilution series should be applied, in petrolatum and aqua. Regarding enzymes however, not the classical procedure leaving the substances for 48 occluded is appropriate, but a fractionated exposure time of 20 minutes, 1 hour, or 4 hours might be important. This is definitely true for ACU, but also for patients suffering from protein contact dermatitis. Dependent on the reactions after certain time-periods, not only the specific diagnosis but also the consequences might differ.

Again, in patients suffering from ACU and even in cases of delayed protein contact dermatitis, positive skin prick tests and detection of IgE-antibodies might be the appropriate test procedure. In some patients, all these tests, patch tests and those relevant for immediate reactions, may cause positive results. As controls, 20 non-exposed subjects (among them 10 atopics) should be tested using the same procedures, with negative results. Currently, there are scanty data on how to patch test with industrial enzymes. Optimal patch test concentrations need to be established for each enzyme, and controls are necessary.

Irritation Caused by Enzyme Exposure

Enzymes, being biologically highly active substances, do exert not only immune-mediated reactions but also effects, that are non-immune mediated.

Respiratory tract irritation

Respiratory tract irritation is a frequent phenomenon caused by many particulate substances and is again dependent on the concentration (dusty environment), the product itself (irritative capacity) and the specific situation of the patient (many asthmatics suffer from bronchial hyperreactivity, where specific (= allergy) but also nonspecific (= irritant) stimuli may cause inflammation).

Skin irritation

Dermal irritation is the production of reversible tissue damage of the skin following the application of a test substance for up to 4 hrs. Skin irritation may occur non-specifically due to dust exposure of sensitive/atopic skin or specifically due to the intrinsic properties of enzymes (proteases and lipases) by digestion of the epidermis. The extent of damage is dependent on the specific activity of the substance (and co-substances, e. g. enzyme plus detergent), but also the time of exposure, and the condition of the skin (normal, damaged, sensitive/damage-prone). Factors like skin-care, cleaning procedures, climatic situation at the place of exposure (heat and humidity) may also play an important role.

Test Methods for Irritative Capacity

Test methods for the irritative capacity of a substance are difficult to define and standardize. Regarding intrinsic bronchial hyperreactivity, nonspecific provocation tests are defined using nonspecific stimuli such as cold air, histamine, codeine, methacholine or others and then perform a lung function test. The irritative capacity of certain occupational substances such as flour, isocyanates, acrylates or epoxices to the respiratory system, is however difficult to determine since well-defined and standardized procedures or recommendations are lacking. Regarding skin irritation, the same problems do exist.

For, among others, cosmetic ingredients and in general chemicals, the OECD Test Guidelines for acute dermal irritation/corrosion and acute eye irritation/ corrosion have just become available in a revised version: Similarly to the testing for the sensitising capacity of a substance, decision trees for testing chemicals were established in the past. The updated version of Guideline 404 includes the recommendation of an integrated strategy, in that an in vivo test should be considered only, if a determination of irritancy or corrosivity cannot be made using structure activity relationship (SAR), in vitro, or other non-animal procedures.

Human studies as last step should only follow the whole cascade, including animal studies, and only if necessary and ethically justified. The OECD method 404 corresponds to EC method B.4 which will be adapted accordingly. In a well-advanced evaluation of a human patch test. as the last step of evaluation, if all previous steps were negative for the identification and classification of a skin irritation potential, results with 29 test materials are presented; all of them being low molecular weight chemicals, with no comments on proteins in general or enzymes in particular. The OECD draft states, that for testing in animals liquid substances are generally used undiluted; regarding enzymes, enzyme concentrates and not enzyme preparations should be tested. No specific test recommendations are established for the determination of the irritative capacity of enzymes in the literature, but testing may follow the OECD guidelines.

Pseudoallergic Reactions

Pseudoallergic reactions (i. e. reactions that clinically imitate allergic reactions) have mostly been reported when enzymes were taken as drugs. Symptoms were skin rashes, fever, arthralgias, myalgias, renal affections, bronchospasm,... and others. Risk-factors for developing allergy: General risk factors for developing allergic symptoms upon contact with a potential allergen are atopic constitution, extent of exposure, and, according to some investigations, smoking (tobacco smoke stimulates IgE-synthesis in animals and might thus increase the chance of sensitisation). There is considerable discussion in the literature, whether persons with atopic constitution or smokers should be excluded from certain allergen-exposing, and thus potentially dangerous, professions, since evidence for smoking being in fact a risk problem (in humans) is only speculation and the atopic constitution frequently difficult to define.

The extent of exposure to (inhaled) antigens however unequivocally correlates with the chance of a person to become sensitised; this is especially true for enzyme antigens. The amount of enzyme concentration in the air thus has to be kept as low as technically possible. There are no reports in the literature on age or sex being specific risk factors. Atopic disorders such as asthma and allergic rhinitis as well as drug allergy are apparently increased in patients infected with human immunodeficiency virus (HIV), but there is no obvious direct correlation between

HIV infection and food allergy. Regarding enzyme allergy and HIV infection, there are no reported cases in the literature.

Detergent Industries

The first adverse effects from handling detergent enzymes were reported in 1969 by Flindt. He observed an epidemic among workers in a detergent powder factory in 1967: 28 patients had symptoms of the respiratory tract. Positive skin-prick tests and precipitating antibodies with detergent enzyme extracts indicated an immediate allergy. PEPYS et al. reported 1969 immunological findings in 3 selected patients referred from Flindt. Apart from reproducing the skin prick tests, they found immediate and late asthmatic reactions to inhalation tests with enzyme powder.

Occupational asthma was also reported in the same year in the Swiss literature. And general industry experience in the early 1970s was based on 3 studies, where sensitisation and respiratory symptoms were measured at 3 sites operated by different detergent producers. In total, the studies covered 1000 workers: 40% were sensitised and 15% exhibited respiratory symptoms. Several cases of enyzme allergy among consumers resulted in the removal of proteases from their products by most manufacturers until methods for enzyme encapsulation were developed and introduced in the 1970s.

Since then the problem seems to have decreased, because industry airborne dust levels were decreased, enzymes encapsulated, safe practices introduced, and older products replaced by antigenically distinct proteases. However, still numerous studies dealing with respiratory allergy in the detergent industry have been published; and continue to be published. And although detergent enzymes are currently unusual causes of occupational allergy, a recent study indicates that they may be more common than reported; and that the development of asthma continues to be a hazard.

Encapsulation alone is not the key to control exposure to enzymes as shown in a modern British factory, which had exclusively used encapsulated enzymes; the rates of sensitisation to enzymes and occupational asthma were probably as high, or even higher, than within the industry before encapsulation. The authors suggested that encapsulation alone is insufficient to prevent enzyme-induced allergy and asthma. The introduction of new enzymes, such as bacterial amylases, and even more modern enzyme mixtures might increase the risk of allergy to amylase and associated asthma in the absence of appropriate preventive measures.

Sensitisations occurred to proteases (esperase and maxapem), lipase, and cellulase. All the sensitised people had symptoms at work; all suffering from rhinitis and 1 from asthma. Summarizing the data from the detergent industry leads to heterogeneous results: A high prevalence (22%) of sensitisation to enzymes was found in a modern and clean factory in a recent study. By comparison, prevalences of 5% to 40% were reported in the detergent industry in the early 1970s. Later, SARLO et al. reported prevalence of sensitisation to be 3.6% - 11.6% during a period of 6 years from 1986 to 1991.

There are several reasons for these discrepancies, among them: enzymes with more potent catalytic activity were introduced, such as lipases and cellulase. And the selection processes for patients are not comparable in the different studies. In some, all exposed workers were investigated, in others it was shown that enzyme sensitisation and work-related respiratory symptoms were positively correlated with airborne enzyme exposure that varies quite considerably within one factory being dependent on the specific workplace.

Pharmaceutical Industry

The discrepancy on reports concerning workers and users is remarkable. Whereas production workers might be affected, reports on problems in users are almost completely missing. Two patients with asthma induced by occupational exposure to cellulase powder derived from Aspergillus niger are reported; cellulase, obtained from a variety of moulds, mainly Trichoderma viride and A. niger cultures, are used as an important member of combined digestive aids or for the removal or softening of unwanted cellulase e. g. in food preparations. 10 sensitised and 10 non-sensitised workers from a pharmaceutical factory who had been exposed to powdered trypsin, chymotrypsin, bromelain, papain, amylase, and lipase were investigated.

Skin test reactivity revealed multiple sensitisations to proteolytic enzymes, i. e. papain (9 out of 10), trypsin (8/10), chymotrypsin (8/10), and bromelain (7/10). It was hypothesized, that the low molecular weight of proteolytic enzymes and their biologic activity might facilitate mucosal penetration much more easily and thus compared to amylase (3/10) and lipase (3/10) permit an immune response and induction of allergic hypersensitivity more easily. Only half of the sensitised workers complained of work-related symptoms. And the fact, that the exchange of four residues in the active site and two non-structural surface loops of trypsin gave the substrate sensitivity of chymotrypsin further substantiates this assumption.

A multisensitised person developed an anaphylactic reaction following the intake of the seventh tablet of Ananase, used as an antiinflammatory drug containing bromelain as the active ingredient. Sensitisation to bromelain usually follows inhalation, and not its ingestion, but cross-reactivity between bromelain and celery, papain, carrot, fennel, and cypress pollen, grass pollen and Compositae have been described. Therefore, the administration of drugs containing bromelain should be carefully evaluated in patients presenting allergic reactions to foods which may cross-react with such enzymes.

Baking Industry

Allergy to flour dust in bakers has been known since 1700. Bakers asthma is a major occupational disease; it has been estimated that 10% of unselected bakers may develop occupational asthma and that 20% of bakers apprentices develop skin sensitivity. Problems are thought to be mainly caused by IgE-mediated sensitisation to cereal mainly wheat, rye and barley proteins. Many allergens have been identified in the water soluble albumin fraction of flour protein, but also in the gluten fractions including globulin, gliadin, and glutenin; mould spores of Aspergillus and Alternaria species, as well as grain mite allergens have been implicated as etiologies of bakers asthma.

And recently enzymes were added to this long list. Enzymes and other flour additives are being increasingly used in bakeries to improve the properties of the baked products and accelerate the baking process. And parallel to this in- creased use, reports on skin and respiratory allergic symptoms were published. α-amylase has long been known to cause occupational (bakers) asthma and rhinitis ; SANDIFORD et al. have shown that beta-amylases are also allergenic. The onset of problems sometimes starts within a few months, sometimes only years after starting to work as baker. The onset of nasal and respiratory symptoms (when sensitised) is within minutes of exposure to α-amylase-treated flour dust and noticeable improvement while away from work during holidays points to the symptoms' occupational nature.

A British study suggests that exposure to α-amylase is a considerable health risk in British bakeries and flour mills, dependent not so much on the exposure time but on peak concentrations. Thus, it is not so much the profession baker itself, that is dangerous, but the specific factory, and in that factory the special places and functions, that cause high or low exposure.

In one study from Sudan, FAKHRI found a high percentage of positive skin test responses to moulds in flour mill workers. In that study, Aspergillus sp and Rhizopus niger were the predominant fungi in the wheat grains. And in a Spanish study, 4.6% of a group of non α-amylase-exposed millers had a positive skin test and 2.7% IgE antibodies. Most of these persons also suffered from clinical problems. The authors conclude, that millers may become sensitised to Aspergillus oryzae or other A species even if never directly exposed (on purpose) to added enzymes.

In a recent extensive study from Finland, 8% of the workers in bakeries (fungal α-amylase and fungal proteases), 5% in the flour mill (fungal α-amylase), and 3% in the crisp bread factory (cellulase, glucose oxidase) were skin test positive to enzymes; the corresponding percentages of positive reactions to flours were 12%, 5%, and 8%. Half of them suffered from allergic symptoms, such as cough, dyspnoea and rhinitis. The risk of asthma is less dependent on the cumulative dose of inhaled dust than on current exposure (at the time of falling ill), and these concentrations vary considerably regarding different bakeries and the special work-places in a certain bakery.

Enzyme Producing Plant

The allergy risk for workers is considerable. In a retrospective study covering the period 1970-1992, 8.8% of the workforce developed clinical enzyme allergy during the first 3 years of employment, half of the cases occurred within the first 15 months of exposure. Sensitisation occurred to all types of enzymes handled in the plant, most often in production areas and laboratories. Ranking diagnoses of enzyme allergy by severity, the frequency of asthma was 5.3%, rhinitis 3%, and urticaria 0.6%. The risk declined during the period of investigation, probably due to improved industrial hygiene and training. But a risk remains.

Some studies report the risk of enzyme allergy in industries using enzymes, especially in production of detergents with enzymes added. From these studies prevalences of 17% - 21% for skin tests were found, and FLOOD et al estimate the incidence proportion of symptoms as 4.5%. These figures cover a period in which the formulation changed from powders to granulates. Since then coating of the granulates further reduced enzyme dust concentration. In certain detergent producing factories using enzymes, respiratory incidence throughout the 1990s have been in the range of 0. 0.03% of the exposed population. This is a 700-fold reduction as compared with general detergent industry's experience of 15% in the early 1970s.

Animal Feed Industry

In this new area of enzyme applications, sensitisation to enzymes was found in 10 out of 140 (7%) of the process workers. as compared to a prevalence of sensitisation of 22% of process workers in a comparable detergent factory, and 20.6% in a group of workers often exposed and 11.8% of workers occasionally exposed in two enzyme producing plants. Only 6 out of these 10 suffered from symptoms at work, mostly rhinitis, but dyspnoea only occasionally. 2 out of the 10 sensitised persons reacted merely to the protease Multifect P 3000, the other 8 reacted to several enzymes: 7 to cellulase, xylanase, or beta-glucanase, 5 to phytase, 3 to glucoamylase, and 1 to α-amylase. 6 of the enzyme positive workers also reacted to environmental allergens, and 4 to flours.

New Types of Enzymes

Any chemical substance may represent a potential allergen. The introduction of new types of enzymes thus poses a potential new risk. New sources of enzymes, enzymes from organisms which were hitherto difficult to handle or non-culturable, genetic engineering techniques and other developments may on the one hand lead to the introduction of new allergens, new (and non-suspected) sources of allergens or exposure to higher concentrations of allergens.

Changes in the amino acid sequence of enzymes, their structure, or properties such as thermostability might change the allergenicity. Thus, in the assessment of the allergenicity of new enzymes., the characteristics of the novel proteins must be evaluated in light of their similarities to known enzymes, food, and environmental allergens. But evaluation should not only be restricted to the rules established for the assessment of the allergenic potential of foods derived from genetically modified crop plants, since that decision tree was developed for ingested but not inhaled proteins. Testing for the allergenic potential of new enzymes should thus follow the regulations as suggested for any new substance.

Safety Evaluation of Enzymes

Any enzyme, representing an existing or a new product, being produced either by conventional or new technologies, theoretically should be tested for its allergenic potential like any other new chemical before it is placed on the market. The criteria for classification of skin- and airway-sensitising substances in the work and general environment were defined only recently. In brief, a substance may be termed being a skin sensitiser (causing contact allergy)

— if studies in at least two independent centres report significant numbers of clinically relevant sensitisations, i.e. association of symptoms and exposure, or

— if epidemiological studies reveal an association of sensitisation and exposure, or

— if case reports from more than one patient from at least 2 independent centres present clinically relevant sensitisation, or

— if at least one positive animal study is published, performed according to relevant test methods without using adjuvants, or

— if at least two not so well documented positive animal studies, one of them performed without adjuvant, showed sensitisation.

A substance may be termed being a proven airway sensitiser (inhalant allergy. i.e. type I allergy, IgE-mediated). If studies or case reports document specific hypersensitivity of the upper airways or the lung, that indicate an immunologic pathomechanism, and stem from more than one patient and from at least two independent centres. An association of exposure and symptoms or functional incapabilities of the upper or lower airways must be proven.

Investigations show that certain enzymes may bear a different potential, dependent on the source example: sensitisation to fungal but not bacterial α-amylases in a baker and their employment example: occupational allergy to fungal α-amylases without relevance for ingested products containing α-amylases. Therefore both, the intrinsic capacity of a certain enzyme to sensitise and the relevance of this capacity, being also dependent on the type of exposure, have to be determined. The safety evaluation of enzymes used in food processing nowadays includes several steps. And based upon a no-effect level derived from in vitro and animal feeding studies it may be concluded that enzymes are safe for the intended use; if all the essential studies are being performed.

Regarding however the allergic potential that an enzyme may exert during production or its employment, additional tests are necessary. In analogy to common chemicals, the Buehler test with guinea pigs was used in some studies. This model likė many other animal models was developed and is validated for the potential of chemicals to cause skin sensitisation (contact allergy) and not for enzymes. It seems evident that proteins are able to cause an immune response through the formation of B-cell mediated antibodies but it has not been assessed to what extent a T-cell mediated type IV reaction is involved. The model is therefore considered not suitable to predict the potential of enzyme preparations to cause allergic contact dermatitis in man and one should therefore rely on human experience.

In an safety evaluation of β-glucanase derived from Trichoderma reesei, a strong skin-sensitising potential of this enzyme was detected using a Magnusson & Kligman guinea pig model. According to the EEC criteria the tested product should be labelled as a skin sensitiser. The data indicate, according to the authors, that there is a potential risk for the workers in the poultry feed industry when handling the preparation. The workers should take precautions to avoid inhalation and skin contact. In the feed compounding in-dustries, this substance is added to barley-based poultry diets in concentrations of approximately 100ppm (0.01%, w/w).

According to the EEC directive on dangerous preparations and substances, preparations containing less than 1% of a sensitiser should not be labelled. The risks of sensitisation are considered to be acceptable when taking into consideration

(a) the strong sensitising potential of the original substance by epidermal exposure,

(b) the normal occupational health precautions taken in the manufacture and use of enzymes,

(c) the extremely low exposure of workers in the feed-compounding industry who take due precautions and

(d) the fact that strong sensitisers may be used safely if there is a short period of contact between a low concentration and a normal skin.

And indeed, the sensitising potential for humans can be judged from test results in guinea pigs only by taking into account the probable human exposure. This exposure may be influenced by several variables, such as the nature of formulation of the enzyme product and the frequency and duration of exposure. These detailed presentations of two controversial ? studies should highlight the current problem regarding studies on the sensitisation potential: Studies on this potential of (low molecular weight) chemicals should be performed in the guinea pig according to OECD 406 and EEC 92/69EEC B6. This is a test for the prediction of contact allergy.

And positive results may or may not be relevant, depending on the time of skin exposure, concentration and formulation of the substance, and many other factors. But there is no recommended test for (high molecular weight) proteins such as enzymes, that normally do not induce contact allergy (type IV reactions) but may cause immediate type, IgE mediated type I reactions. One author regards the enzymes under investigation not as skin-sensitiser despite a positive Buehler test, the other concludes from positive Magnusson-Kligman tests (and other test systems) that the enzyme is safe for use in feed of the intended target species; that, however, some occupational health precautions should be taken to avoid skin contact and inhalation, as is the case for almost all enzyme proteins..

And indeed, test systems were only recently introduced to study the capability of enzymes to induce antibody responses: SARLO et al. studied the influence of MHC background on the antibody response to detergent enzymes in the mouse intranasal test. The authors could demonstrate, that different enzymes induced antibody responses to a different extent, and this reaction was dependent on many factors. The hybrid strain BDF1 ranks enzymes the same as the guinea pig, which in turn correlates with sensitisation in occupationally exposed humans. The ranking was based upon the dose of enzyme needed to give one-half maximal IgG1 antibody response (ED-50) where Termamyl is more potent than Alcalase, which is equipotent to Savinase.

And BARTON et al. could demonstrate that four different proteases caused differential effects dependent on the immunization system, the kind of animal and the enzyme itself (subtilisin B > pancreatin, and subtilisin A > papain). Animal models were developed to evaluate all new enzymes for their potential to induce the formation of allergic antibodies recently. In the guinea-pig intratracheal test (GPIT) animals were exposed to defined protein levels of enzyme once/week for 10-12 weeks. Sera were collected at selected time-points and the amount of enzyme-specific IgG1 homocytotropic antibody measured by either passive cutaneous anaphylaxis test or by ELISA.

A reference or benchmark enzyme, Alcalase (subtilisin), was always included in the test. And in 1996, ROBINSON et al. developed a mouse model as an alternative to the guinea-pig. Intranasal exposure of BDF1 mice to enzymes induced a dose-dependent and enzyme specific IgG1 antibody response after the exposure.

Extension of the dosing regimen led to the generation of enzyme specific IgE antibodies.

But: many questions remain, a generally accepted system to predict allergenicity of enzyme preparations is not available, and key questions remain unanswered like: the clinical relevance of current animal and in vitro test systems, the comparability of in vitro and in vivo assays, and probably quite important regarding the hazard of enzymes: how relevant is the dose level versus duration of exposure? e. g. high dose acute or subchronic exposure vs. low dose chronic exposure. In addition, co-factors such as surfactants present in detergent products have been reported to exert adjuvant activity, that make literature data difficult to compare.

Evidence is abundant, that many enzymes can induce immediate type reactions in man upon relevant exposure. Since neither safe limits for exposure time nor concentration could be defined yet for any enzyme, any exposure should be kept as limited as possible. And in fact, while in previous reports a safe exposure limit of 60 ng/m³ has been recommended, many studies have shown that sensitisation and symptomatic allergy may occur at much lower levels. Even for job titles with a relatively low exposure to fungal α-amylase of less than 2-5 ng/m³, and in some cases even in a concentration below the detection level of the assay (0.3-0.5 ng/m³), a clearly enhanced risk of sensitisation was demonstrated.

If a safe occupational exposure limit for fungal α-amylase can be defined, it should be around or below the detection limit of the assay, and far below the previously recommended value for enzyme dust (60 ng/m³). And this may not only be true for fungal α-amylase but for all enzymes. at least until safe limits are proven for a certain preparation. The low molecular weight of proteolytic enzymes and their biologic activity might facilitate mucosal penetration more easily and thus compared to amylase and lipase permit an immune response and induction of allergic hypersensitivity more likely amylases glucanase.

Negative tests for skin sensitisation should be taken with care; in the safety evaluation of a lipase expressed in Aspergillus oryzae, in the first epicutaneous challenge positive responses were seen in three of 20 test animals; for several reasons it was concluded that sensitisation had not been shown with the lipase. Regarding new enzymes, a decision tree approach for the evaluation of their allergenicity is mandatory in a similar way as it has become obligatory for genetically modified foods. This includes,

— animal models

— amino acid sequence comparison with known allergens

— in vitro and in vivo tests in allergic and non-allergic individuals

— testing for physico-chemical characteristics (common food allergens are resistant to proteolysis, stability to digestion in the gastrointestinal tract, have a tendency to be abundant, are often glycosylated, and be resistant to food processing or cooking

— adequate post marketing surveillance.

Highly allergenic proteins are often expressed at relatively high levels. However, allergens can sensitise susceptible individuals at less than milligram levels, possibly at less than microgram levels. The elicitation of objective symptoms in already sensitised individuals can also occur at low levels of exposure, but has not been documented below 500 micrograms. It is therefore not possible to define a level of expression below which a protein can be considered safe from the allergenicity point of view. Thus, currently level of expression cannot be incorporated into the assessment of the allergenicity of genetically modified foods.

Safety Evaluation of Enzymes

The regulator should define and validate the test systems to be used, in vitro and in vivo (animal studies). Currently, negative skin-responses associated with repetitive, low-dose exposure to industrial chemicals and consumer products all too often are not accurately predicted by the required assays. Predictive assays generally require data on the dose-response relationship for prudent and optimal interpretation.

There are many critical variables which, when data are available, can facilitate more robust and sound information. Currently several steps of testing are in use for the evaluation of the allergenic potential of chemicals, but all these studies may predict a potential for the induction of contact allergy by chemicals only. None is capable of exactly predicting immediate type reactions; and numerous artefacts may lead to either false-positive or false-negative results.

Regarding predicitive assays for evaluating the ability of enzymes to produce immunological, immediate type I allergic reactions, test methods are scarce, controversial, and of limited clinical relevance, hardly standardized and exclusively based on animal models. The recently published mouse intranasal test, developed to assess the immunogenic potential of detergent enzymes, uses inbred mouse strains of different haplotypes, hybrid and congenic strains. Certain strains ranked enzymes the same as the guinea pig, which in turn correlates with sensitisation in occupationally exposed humans. Studies in mice have also demonstrated, that enzymatically active papain preferentially induces an IgG1 response and results in mast cell degranulation, both features typical of an allergic reaction.

In contrast, inactive papain desensitised mice to subsequent challenge by active enzyme. Is is only the recently developed guinea-pig intratracheal test and the mouse-intranasal exposure test, that seem to produce reproducible results in the respective systems. These examples may demonstrate, that progress is being made by adapting techniques for the predictive study of enzymes to induce or elicit immediate type reactions in animals. The question on the relevance for man remains to be answered. And time does not seem ready to recommend certain test methods routinely for such evaluations yet. In addition, the potential for interaction among different enzymes with an effect on sensitisation as well as adjuvant factors important for an increased immune responses remain to be uniformly defined.

Protection of Workers

Occupational asthma and allergy associated with the use of enzymes was first reported soon after the introduction of alkaline and heat stable proteolytic enzymes into detergent products in the 1960s. But despite encapsulation of the enzymes and improved hygiene at manufacturing sites, the development of asthma has continued to be a hazard. It was only recently that effective control programs in reducing exposure, based on a surveillance programme to identify and correct excess exposures before symptomatic illness occurs, that numbers of new cases have declined. This approach includes coating of enzymes, compliance with operational guidelines, introduction of exposure guidelines for airborne enzyme proteins, reduction of peak exposures, training, medical monitoring, commitment and changes of behaviour, surveillance systems, etc..

Manufacturing sites of enzymes should adhere to operational guidelines that reflect daily safe work practices for controlling dust and enzyme exposure. These include: no visible dust, no recurring spills and temporary repairs, no gross skin contact and the treatment of empty containers as if full of enzyme products. These guidelines are met by engineering controls to minimize dust and aerosol generation during normal operation and peak exposure situations, product formulations that minimize dust and aerosol generation (encapsulation by inert materials), limits on the level of enzyme in detergent product, air monitoring systems to verify adherence to exposure guidelines and employee education and training to minimize personal exposure.

Ventilation systems must adequately filter and not recirculate particulate matters. Companies must fully comply with OSHA standards. In the baking industry, where the use of enzymes is likely to continue and increase, less dusty preparations, such as granules or oil suspensions, should be introduced. Exhaust hoods and local ventilation systems especially in the dusty phases, like the weighing processes, and on the dough pots are needed. Replacement of fungal α-amylase by bacterial α-amylase should be considered, since no reactions were found to the latter, indicating different antigenic characteristics, which were confirmed by immunoblotting and RAST inhibition.

Measuring wheat flour specific IgG antibodies as an index of allergen exposure may help monitor this exposure, since there was a tendency for IgE and IgG1 levels to decrease after the installation of an air purifier and the introduction of other protective measures. Workers handling enzymes should use protective clothing and eye protection. Water should be easily accessible in case spillage or splashing occurs. Adequate protective breathing equipment should be used if dust or aerosols are encountered. Skin exposed to enzymes should be washed immediately, and eyewash and shower stations must be convenient and fully functional.

If enzyme is inadvertently swallowed, adequate rinsing and water should be given, and medical advice should be sought immediately. Plant and laboratory practices should emphasize safety and good housekeeping. Enzymes can be allergenic, depending on the susceptibility of individuals. Adequate precautions to protect the individual include optimal skin care, where skin contact cannot be avoided as in bakeries, since eczematous skin is prone to become sensitised more easily.

Despite enzyme encapsulation and modern process techniques in the detergent industry, there still seems to be a risk of allergy. Exposure levels may vary widely, probably from >10μg/m³ at sites of primary production and packaging of purified preparations, to less than 1 ng/m³ where enzyme preparations are finally used, like in bakeries. Air sampling may be performed by static samplers, distributed at different working places, or by individual sampling, where workers wear a personal sampler for a whole shift. Enzyme concentration measurements at worksites have mainly focused on protease production and its use in the detergent industry.

It is not known which concentrations are capable of sensitising workers, but it is probable that concentrations below the threshold limit value (TLV) proposed by the American Conference of Governmental Industrial Hygienists of 60 ng/m³ can sensitise (ACGIH), or at least cause symptoms in sensitised workers. This guideline was based on the protein content and activity of enzyme receipt materials, estimation of exposure levels during the late 1960s and measurement of enzyme levels in facilities that had eliminated asthma and allergy symptoms. The detergent industry moved to levels lower than the ACGIH TLV due to the suspected potentiation of allergy to enzymes by detergent matrix. Thus some detergent companies use an internal occupational exposure guideline of 15 ng protein/m³.

Protease concentrations were generally < 20 ng/m³ in a Finish detergent factory, but occasional peak values up to 80 ng/m³ were detected in the packing and maintenance tasks, and high values of > 1 μg/m³ in the mixing area. In a study on fungal α-amylase allergy in bakery workers, a dose-response relation between the job title-related mean exposure level assessed by personal airborne dust sampling combined with a sensitive enzyme immunoassay for α-amylase and both sensitisation and the presence of respiratory symptoms were evaluated.

Even for job titles with a relatively low exposure of less than 2-5ng/m³, and in many cases a concentration below the detection level of the assay (0.3-0.5 ng/m³), a clearly enhanced risk of sensitisation was demonstrated. The authors conclude, that a safe occupational exposure limit for fungal α-amylase should be around or below the detection limit of the assay, and far below the previously recommended value for enzyme dust. Personal sampling revealed that a small proportion of workers in bakeries are exposed to α-amylase at concentrations that result in high rates of sensitisation; and exposure to α-amylase showed only a moderate correlation with concentrations of dust in these workers.

Finland specifies no hygienic limit values for occupational exposure to any enzyme. The occupational limit of exposure to organic dust is 5 mg/m³ (eight hour average) and 10 mg/m³ (15 minute average). Total measured dust concentrations in bakeries were from 0.1 to 18 mg/m³ (thus exceeding the limits by far), with highest values in dough making areas. α-amylase generally followed the total dust concentrations and reached the highest values of < 6.6μg/m³ in the same areas. Cellulase and xylanase varied with concentrations < 180 ng/m³ and < 40ng/m³, respectively, in the flour mill and the crispbread factory. No cellulase, but concentrations of 1-200ng/m³ xylanase, were found in the bakeries, probably indicating the natural xylanase activity of wheat.

In the animal feed industry, enzyme concentrations in the air varied greatly: xylanase from less than 0.8 ng/m³ up to 16 ng/m³, α-amylase from less than 20 ng/m³ up to 200 ng/m³, and protease from less than 0.4 ng/m³ up to 2900 ng/m³. Exposure to enzymes does not uniformly follow exposure to total dust: added enzymes are not present in the grain when it is received, and their concentration is low in the end product. Between these stages, exposure to enzymes is possible in the adding of premixes and other process phases. Several studies point to the importance of the assay method: Some methods are based on the enzyme activity; thus only active enzyme is measured, omitting inactive but potentially sensitising enzyme proteins.

In addition, methods may not be specific for the origin of the enzyme activity, as would be an immunological assay: colorimetric assays for proteases may detect microbial enzymes as well as proteolytic activity of the grain dust. Flour, especially wheat, contains natural α-amylase. But the extent to which the measured α-amylase in air is due to flours cannot be assessed with colorimetric methods. Contamination exerts another problem when controlling dust exposure in bakeries. The industrial grades of enzymes are poorly purified, they contain varying amounts of other enzymes produced by the microorganism in question. Thus α-amylase, for example, may contain proteases, and cellulases xylanases, and vice versa. This may cause unexpected sensitisations and cross reactions.

Several studies have clearly demonstrated, that not so much the amount of exposure over time, but exposure peaks exceeding the mean concentrations, do represent the major problem for sensitisation. Since however the exposure levels are usually mean values for job titles and production facilities, it is not clear whether sensitisation in a medium exposure population was due to this mean or a cumulative dose of allergen, or to repeated short peak exposures, e. g. of 10-100 ng/m^3 or more, which most probably occur with some frequency. It thus might be speculated that time-weighted average levels of more than 5-10 ng/m^3 could be 'safe., if such peak exposures can be effectively prevented.

However, as long as no experimental or observational data are available to resolve this question, it seems reasonable to use time-weighted average values as the basis for any exposure limit. As shown in recent reports by the industry, it seems practicable to reduce enzyme concentrations far below the TLV. The control of occasional peak exposures, which are probably important in inducing sensitisation, remains a principal challenge. Indirectly controlling total exposure over time by monitoring the patients antigen-specific IgG antibodies may at least in the baking industry be a feasible task.

6

Health Foods and Medical Biotechnology

Novel biotechnology research is rapidly expanding with a view to the application of environment friendly bioprocesses for chemical, pharmaceutical, and other areas, including food bioengineering and safety. The tremendous growth of the two areas of modern biotechnology may have recently created the impression of technological isolation through specialization of both agricultural and medical biotechnology. Such an impression-if not checked from spreading- might raise artificial barriers between these two important and closely related domains.

Caused mainly by over-specialization, an increasingly 'narrow' approach to biotechnology might affect the evolution and sharing of existing biotechnological tools that were initially developed for different types of applications. For example, the current adoption of certain genetically modified foods and the administration of vaccines based on genetically modified organisms (GMO's) in developing countries, EU and Japan raise distribution problems for which adequate solutions have yet to be found. Potential loss of certain crop markets (such as corn) that may have included 'GMO-mixed' crops could be very substantial (>1 billion US $).

Therefore, solutions to such problems must, and can be, found both through crop growing policy changes and also by utilizing novel, ultra-sensitive, as well as less expensive, techniques for GMO detection and crop quality control. Unfortunately, about one billion people in the world remain deficient in several vitamins and minerals intake. Such undernourished populations live mostly in developing countries that often have inhospitable climates which favor low efficiency in food engineering and production, and also where the primary services for storage or administration of GMO-based vaccines are minimal.

The utilization-in combination with agronomical techniques-of biotechnological tools that were originally produced for the medical area, may help resolve such difficult problems. As an example, the production of vaccines from plants and microorganism with subsequent addition in foods could reduce the cost of vaccination in such areas. The modification of microorganisms for the bio-production of fuels is also important, and would also be important in under-developed, or developing, countries that have a large surplus of under-utilized complex carbohydrate sources that could be used for bioproduction of valuable fuels with existing technologies.

Immobilisation of Cells

Immobilization of whole cells has been defined as the physical confinement or localization of intact cells to a certain defined region of space with preservation of some, or most, catalytic activity. The increased stability under extreme conditions of pH and temperature, as well as the re-use and applicability in continuous processing systems that enclose immobilized cells instead of soluble enzymes make the cells a preferred, versatile tool in both food industry and medicine. There are several different approaches to the classification of immobilized biocatalysts, but the most frequently employed classification is based upon the method of immobilization selected for a specific application.

The selection of immobilization method depends, therefore, upon the application, the nature of the microorganism being immobilized, as well as the resources available. Some of the advantages of whole-cell immobilization in comparison with enzyme immobilization are: the higher stability and enzyme activity, multivariate enzyme applications, and the lower cost. On the other hand, disadvantages of using whole cell immobilization in comparison with enzyme immobilization are linked to the increased diffusional barriers caused by the much larger sizes of cells in comparison with enzymes.

Adsorption is the least expensive and mildest immobilization method. It uses weak interaction forces such as hydrogen bonds, hydrophobic interactions and Van der Waal forces to immobilize cells or enzymes. However, the sensitivity of this interaction to pH makes the leakage of cells immobilized by this technique quite common. Important applications of this technique are related to the production of fructose and vinegar, and also waste water treatments. Ionic binding uses the properties of negatively charged microbial cells to interact with positively charged ion exchangers. The results obtained with this technique are also sensitive to extreme pH values, and the binding strength is greater in comparison with adsorption.

However, the mild conditions employed by this technique make it suitable for use for immobilization of both enzymes and whole cells. Covalent binding and cross-linking offer better strength than the previous techniques, however, there is an encountered toxicity in the reagents that are used to produce immobilization. Entrapment techniques are, however, the most commonly encountered in the industry and they are based on the formation of thermally reversible gels, ionotropic gels and polymerization.

Food Industry and Human Nutrition

Entrapment of cells in a gel-like matrix by ionotropic gelation using alginates and Kcarrageenans is certainly the most useful method for industrial purposes. The properties of the gel-like matrix allow the cell to remain viable and with its catalytic ability for a long period of time. For example, an increment in the yeast concentration obtained through immobilization techniques has helped the brewing industry to reduce fermentation process times and the size of their storage facilities.

Unfortunately, because of the high concentration of diacetyl, and the low concentration of higher alcohols and esters, the flavor of the fast fermented beer has been compromised. Even the amino acid profile has been altered. The main factor causing this uncommon imbalance is the insufficient mass transfer in the older designs of fermentation reactors; thus, the use of new reactor designs, and combining technologies could improve the quality of the products obtained through such fermentation reactions.

Nutraceuticals are defined as food components that have health benefits beyond traditional nutritional value. Novel biotechnology tools like immobilization were also applied for the isolation and incorporation of such food components in ordinary foods. The synthesis of nutraceuticals was reported to be successful by empoloying immobilized lipases, such as those from Candida antartica and Lactobacillus ruteri. The introduction of conjugated linoleic acid (CLA) in dairy foods has been made possible through the immobilization of lipases.

Immobilization Technique Applications in Medicine

There is a quite extensive list of immobilization technique applications in medicine. A very important group of such applications is concerned with the regulation of equilibrium between coagulation and dissolution of coagulated blood (fibrinolysis) through the use of immobilized enzymes. The high probability of death caused by thrombosis, has committed physicians to the use of fibrinolytic therapy for the treatment of occlusions in those parts of the body where a surgical intervention would be too risky. Amongst the most important enzymes that have been immobilized for use in such therapy are Plasmin and Heparin. The use of biotechnology as well as microscopic techniques has helped refine and greatly improve such therapeutical means.

Current regulations for the disposal of toxic chemicals in the environment as well as the detoxification of water used in any agricultural and industrial process brings the need for novel biotechnology tools to be developed in order to solve such problems in a cost-efficient manner. Enzymes have been isolated from genetically manipulated microorganism strains with the purpose of accelerating the rate of degradation of organic and some inorganic compounds in wastewater as well as in soils.

Genetic Manipulation Techniques

Genetic manipulation techniques have played an important role in enhancing the performance of microorganisms that are significant in the industrial, pharmaceutical and medical application areas. Mutagenesis can be defined as the group of techniques or processes that produce changes in a DNA sequence which modifies either the expression of genes or the structure of the gene products. Two subcategories of mutagenesis are usually applied in the genetic manipulation of microorganisms: classical mutagenesis and transposon-directed mutagenesis. Classical Mutagenesis involves the use of chemical mutagens to regulate a process of interest in the target microorganism.

The main purpose of this 'classical' approach is to find the mechanism of action of the targetted microorganism. Transposon-directed Mutagenesis consists in the genetic alteration of a few sites on the target chromosome in order to determine the function of this specific site of the chromosome. By comparison with the previous category transponson- directed mutagenesis it is more specific, and is therefore preferred in specific applications that require higher selectivity.

Gene Transfer

Electroporation involves the production of transient pores in a bacterial cell membrane following the application of high-voltage electric field (DC) pulses of short duration. Such pores allow the introduction of DNA into the cell under certain, favorable conditions. Such methods have been used with various microorganisms and even plasmids that are of significant use in the industry.

Another technique that is employed for the transference of genes involves Shuttle Vectors. Shuttle vectors are DNA constructs that are able to replicate and deliver DNA to widely divergent types of bacteria. DNA transference is obtained after the use of a microorganism whose mechanisms of action have been thoroughly studied, as for example, is the case of various E. coli strains.

Gene Cloning

Gene cloning and genomic maps to clone the genes of industrially-significant microorganisms have also been reported to be subsequently introduced in native strains of the microorganism of interest.

DNA Analysis

DNA analysis involves the study of a particular region of DNA by either physical or chemical techniques. It is usually carried out in order to obtain insights into the genetic organization, the mechanisms of gene expression, or to obtain information that may be useful for the construction of shuttle vectors.

Microencapsulation

Microencapsulation is now being extensively used in the industry and medicine to enhance the stability and shelf life of products, to improve the nutritional value of foods, and to produce enzyme, or immuno-protective, capsules for the production of vaccines and for the transplant of organs. However, the prospects of microencapsulation techniques are not limited to these areas because it can be used in combination with other novel biotechnology tools to broaden the spectrum of its applications.

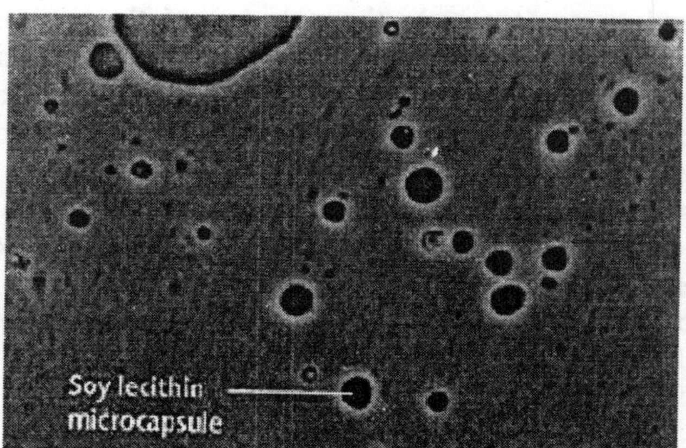

Figure 1. Photomicrograph of Soy Lecithin microcapsules of about one micron in size.

The concept behind microencapsulation involves surrounding a component of interest (such as vitamins, proteins, flavor, etc.) that need protecting with a (µm-thick) thin layer of either polysaccharide or lipid material in order to prevent, or significantly reduce, the deterioration of the encapsulated component. As a result, liquid droplets, solid particles or gaseous materials are packaged with continuous shells that can release their contents at controlled rates under desired conditions.

There are several, 'classical' techniques that are used for microencapsulation such as: spray drying, molecular inclusion and extrusion. On the other hand, liposome microencapsulation is a novel technique that is being developed to the point that it can introduce any substance of interest in an organism regardless of the substance's solubility, electrical charge, molecular size or structure.

Liposome Microencapsulation

A liposome can be defined as an artificial lipid vesicle that has a bilayer phospholipid arrangement with the head groups oriented towards the interior of the bilayer and the acyl group towards the exterior of the membrane facing water (Fig.1). Liposomes are usully made of phosphatidylcholine (lipid) molecules although mixtures of phospholipids can also be employed to make liposomes. The main difference between liposomes and cell membranes is the lack of bilayer asymmetry in the case of liposomes. On the other hand, liposomes do exhibit barrier properties very similar to those of cellular lipid membranes, and therefore, they are frequently used as a model system for the study of the transport activity of the cell membrane boundary.

Since the invention of liposomes in the early 1960's as 'membrane-like' structures, they have been extensively used in the biotechnology field. The 'membrane-like' characteristics, their colloidal size as well as their solubility properties for various molecules have made liposomes useful tools for various biotechnology applications, from foods applications (such as in enzyme microencapsulation), to fermentation and several medical/ pharmaceutical applications.

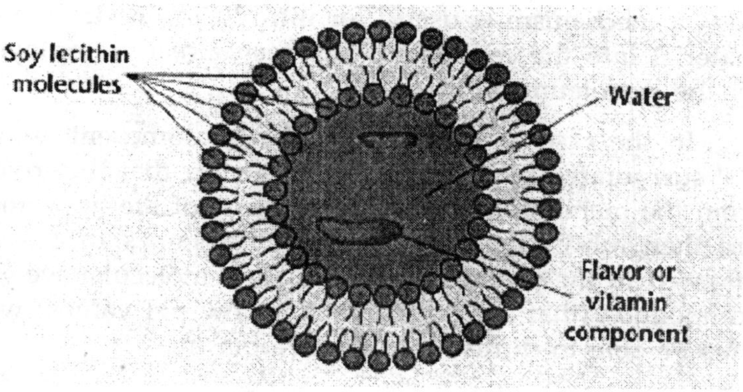

Figure 2. Simplified representation of the molecular organization of a liposome microcapsule in water.

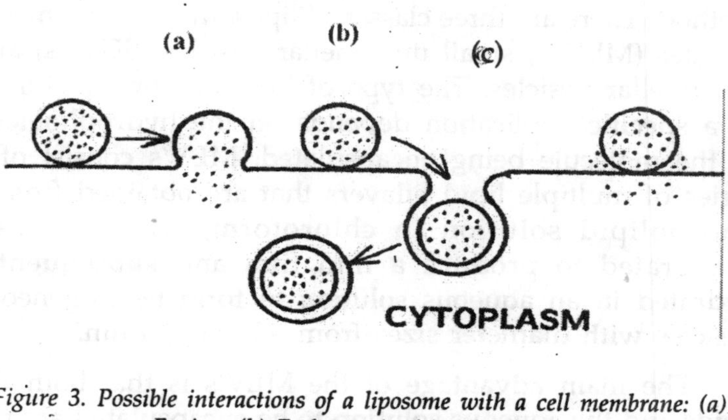

Figure 3. Possible interactions of a liposome with a cell membrane: (a) Fusion; (b) Endocytosis; (c) Adsorption.

A few possible mechanisms for liposome absorption into cells are illustrated in Fig 3. Fusion and endocytosis are the most common ones being utilized for the production of liposomes with industrial or medical purposes. During fusion, the soluble material carried in the liposome is released into the cytoplasm of the target cell through the joining of the liposome with the cell membrane lipids. Such a mechanism is useful for introducing hydrophobic material into the cell, and it has been already employed by the petroleum industry.

In the case of endocytosis, the liposome enters the cell surrounded by an endocytotic vacuole that is derived from the membrane lipids. The contents of the liposomes are then released by the action of liposomes that were attached to the produced vacuole. Low molecular weight compounds (that are not charged at low pH values) will thus be able to diffuse into the cytoplasm.

Classification of Liposomes

Liposomes are classified according to their size and structure, the latter also depends on the preparation method. There are three classes of liposomes: multilamellar vesicles (MLV's), small unilamellar vesicles (SUV' s) and unilamellar vesicles. The type of liposome being utilized in a specific application depends on the hydrophobicity of the molecule being encapsulated. MLV's consist of a series of multiple lipid bilayers that are obtained from a phospolipid solution in chloroform, which is then evaporated to produce a thin film and subsequently hydrated in an aqueous solution to form heterogeneous vesicles with diameter sizes from ~0.3 to 2.0 μm.

The main advantage of the MLV's is that both the lipids and the aqueous solution to be encapsulated are not subjected to harsh treatment. The use of high-intensity ultrasonication, ethanol injection and pressure applied to multillamelar vesicles allows the production of single bilayer vesicles SUV's; this physical treatment reducesthe liposome size. Unfortunately, the smaller size of such a vesicle also results in a higher curvature that limits its capture volume. LUV's are prepared commonly by infusion, reverse-phase evaporation and detergent dilution methods. They vary in size range from 100 -500 nm and they are the most widely employed liposomes because they have less variable size and higher entrapment volumes (>~2.7 L / mol of lipid) than SUV's.

Liposome Microencapsulation Techniques

The lipid-aqueous system needs to meet two major requirements for the liposome microencapsulation to occurr. First, the system needs have a negative free energy, and second, it needs to overcome the energy barrier which is necessary for the formation of the bilayer. Three methods of liposome preparation are here described:

Microfluidization

Microencapsulation by this method is obtained through the dynamic interaction of two pressurized aqueous-lipid fluids that create a large momentum and flow turbulence that allows the system to overcome the energy barrier to microcapsule formation. The pressure applied in air-driven microfluidizers can be as high as 10,000 psi. The ultra-high velocities reached by this technique allow the creation of small liposomes (< ~0.3μm) with high capture efficiency. This system is useful because of its capability to produce very large amounts of liposomes with adjusted size in a continuous process.

Ultrasonication

The ultrasound absorption is employed to overcome the energy barrier. The sonication of the lipid emulsion can be carried out by two different approaches. The first one is through the use of a sonication probe placed directly into the suspension of liposomes. The second method is slower than the first one and employs a sonication bath, such as a sealed container filled with Nitrogen or Argon gas. Both methods have been extensively applied for the formation of SUV' s; however, the use of a sonication probe has been found to cause contamination of the liposomes.

Reverse phase evaporation

Reverse phase evaporation is used for the preparation of LUV's and it is based upon the extraction of a nonpolar solvent from an aqueous-nonpolar inverted micelle by rotatory evaporation under vacuum. This withdrawal of the nonpolar phase changes the intermediate gel-like phase of the micelle into uni-lamellar and oligo-lamellar vesicles. The advantage of this technique is the uniformity of the vesicles formed (from about 0.2 to1.0 μm) as well as their high encapsulation efficiency. On the other hand, the exposure of the components to organic solvents and sonication is likely to result in protein denaturation.

Liposome in Medical Biotechnology

Traditionally, liposomes have been used in the bioengineering field to over-produce certain proteins through the genetic modification of cells. They have thus solved the inconvenience of transferring high molecular weight molecules through cell membranes. The use of unilamelar vesicules made of cationic lipids has improved the transfection efficiencies and prevented interactions with DNA molecules. Therefore, the project of introducing genes to cure diseases (genetic therapy) is not far from becoming a reality if the patient being treated were found not to suffer from severe side effects.

Vaccine formulations based on liposomes have been successfully tested in animal immunization, and such studies are currently in the clinical testing phase. The benefits and limitiations of liposomes as drug carriers in a system as complex as the human body depend basically on the interactions of liposomes with the cells, as well as their immuno-compatibility, or their ability to escape detection by the human immune system.

Cosmetics is another area where the liposomes have been extensively employed. They are being utilized as humectants, as well as carriers of formulations containing extracts, vitamins, moisturizers, antibiotics and proteins. Such applications are mostly directed towards preventing, or delaying, the aging of the skin. Through their surfactant action liposomes also improve the coagulation and sinking of oil spreading at a water interface, a methodology which has been under study for some time by EPA for cleaning up oil spills.

Applications of Lectin

Lectins are proteins, or glycoproteins, that agglutinate erythrocytes of some or all blood groups in vitro. They are an important group of bioactive proteins and glycoproteins found in most organisms, including plants, vertebrates, invertebrates, bacteria and viruses, and have several important applications to the fields of health food and medical biotechnology. Lectins are used as tools in the fields of biochemistry, cell biology and immunology, as well as for diagnostic and therapeutic purposes in cancer research. Lectin aggregation can be employed on a large-scale basis for the commercial production of biologically active proteins.

Lectins have been used in glycoprotein purification, oligosaccharide analysis, as well as in cell-selection processes. Lectins can bind reversibly with free sugars or with sugar residues of polysaccharides, glycoproteins or glycolipids. There has been increasing demand for novel diagnostic and medical cancer therapies that utilize non-traditional sources. Epidemiological studies indicate that the consumption of a plant-based diet is strongly associated with a reduced risk of developing several types of cancer.

Plants contain numerous phytochemicals that can alter cancer-associated biochemical pathways. One such group being intensively examined for its role in cancer chemoprevention is lectins. Lectins are currently being considered for use as cancer therapeutic agents. Lectins were reported to bind preferentially to cancer cell membranes, or their receptors, causing cytotoxicity, apoptosis and/or tumor growth inhibition. Lectins were thought to become internalized into cells, and some lectins were claimed to cause cancer cell agglutination and/or aggregation.

Present in common foods, some lectins resist acid and/or enzymatic digestion and also were reported to enter the bloodstream in an intact, and biologically active, form. Lectins possess a spectrum of beneficial, as well as harmful, effects both in vitro and in vivo. Ingestion of lectins also sequesters the available body pool of polyamines, thereby claimed to thwart cancer cell growth. They have also been reported to affect the immune system by altering the production of various interleukins, or by activating certain protein kinases. Lectins were also reported to bind to ribosomes and thus inhibit protein synthesis.

Lectins may also modify the cell cycle by inducing non-apoptotic G1- phase accumulation mechanisms, G2/M phase cell cycle arrest, and apoptosis, and might activate the caspase cascade. Lectins were also reported to down-regulate telomerase activity and inhibit angiogenesis.

Lectins could inhibit cell adhesion, proliferation, colony formation and hemagglutination, and were reported to have cytotoxic effects on human tumor cells. Lectins could function as surface markers for tumor cell recognition, cell adhesion, signal transduction across the membrane, mitogenic cytotoxicity and apoptosis. Also, lectins were reported to modulate the growth, proliferation and apoptosis of premalignant and malignant cells both in vitro and in vivo. Most of these effects are thought to be mediated by specific cell surface receptors.

For many years, lectins have been considered toxic substances to both cells and animals, mainly because of the observed agglutination of erythrocytes and other cells in vitro. On the other hand, it has also been reported that lectins have an inhibitory effect on the growth of tumors. Their potential for clinical applications has been investigated only in recent years. Lectins are now being considered for use both in the diagnostics and therapeutics of cancers. Thus, lectins are quite versatile biomarkers and have been utilized in a variety of studies involving histochemical, biochemical and functional techniques for cancer cell characterization.

Lectins may also be very useful tools for the identification of cancers and the degree of metastasis, or cancer development stage. Recently, there has been a tendency to shift lectins use from cancer detection to actual use in combating cancer. The reason for this shift is mainly caused by recent research that indicated the cytotoxic and apoptosis/necrosis-inducing effects of certain lectins, combined with the hypothesis that dietary lectins enter the systemic circulation intact. One important feature appears to be that lectins stimulate the human immune system.

Lectins were thus reported to exhibit antitumor and anticarcinogenic activities that could be of substantial benefit in cancer treatment. Extracts of Viscum album (mistletoe) are widely used as complementary cancer therapies in Europe. Mistletoe has been used parenterally for more than 80 years as an anticancer agent with strong immuno-modulating action. The quality of life of patients with late-stage pancreatic cancer was reported to be improved on account of exposure to mistletoe lectin. Immuno-modulation using recombinant ML was reported to influence tumor growth in breast cancer patients.

Bladder carcinoma was reported to be significantly reduced, and survival times were reported to be prolonged in mice as the concentration of ML was increased from 3 to 30 ng. ML increased the life span, decreased the tumor growth and decreased hyperplasia of mice and rats with lymphoma and lung cancer. A lectin purified from mesquite seed was reported to have an anti- proliferative effect on the cervical human tumor cells and on cell adhesion. Interestingly, mesquite lectin modulated the growth, proliferation and apoptosis of HeLa cells, while having no effect on normal cells in vitro. Vicia faba agglutinin (VFA), a lectin from broad beans was reported to aggregate, stimulate the morphological differentiation of, and reduce the malignant phenotype of colon cancer cells.

Wheat germ agglutinin (WGA) proved to be highly toxic to human pancreatic carcinoma cells in vitro. WGA exposure induced chromatin condensation, nuclear fragmentation and DNA release, consistent with apoptosis. The binding of the snail lectin Helix pomatia agglutinin (HPA), which recognizes N-acetylgalactosamine and N-acetylglucosamine sugars, is considered to be a strong predictor of metastasis and unfavorable prognosis in a number human adenocarcinomas, including breast cancer. Because of their carbohydrate bio-recognition properties, lectins may also be used as carriers for targeted drug delivery, in a manner similar to liposomes, provided the possible side effects of such treatments could be minimized.

It has been observed that mucosal expression of terminal unsubstituted galactose is increased colon pre-cancerous conditions and cancer, and that it allows interaction with mitogenic galactose-binding lectins of dietary or microbial origin. Based on this observation, an interesting hypothesis was postulated whereby galactose might be able to prevent cancer by binding and inhibiting such lectins from interacting with colon cancer cells. D-galactose treatment was reported to be effective in liver lectin blocking to prevent hepatic metastases in colorectal carcinoma patients. Epithelial cancer cells showed an increased cell surface expression of mucin antigens with aberrant O-glycosylation, notably Thomsen-Friedenreich Antigens (TFA).

TFA is a carbohydrate antigen with a proven link to malignancy. Immunoassays could be utilized for antigens such as TFA in order to determine the metastatic potential of breast and colon cancer cells. Molecular changes in the membrane surface in the case of both stomach and colon cancer cells occur during the progression to carcinogenesis. Carbohydrate patterns displayed on the cellular membrane exterior are molecular signatures with unique biological characteristics related to oncogenesis and metastasis, and could be used to determine the appropriate chemo-therapeutic and surgical procedures for each specific cancer. Lectins have already a demonstrated potential for the treatment, prevention and diagnosis of chronic diseases such as cancer.

Molecular Modeling

Computational Biology has a very wide range of applications currently thought of 'belonging' to Biotechnology, such as Bioinformatics, even though the development of such computations has preceeded modern Biotechnology by many decades.

Applications of Molecular Modeling

Molecular modeling has been introduced for more than two decades ago. Increasingly, modeling software is available for a variety of industrial applications. Global markets for molecular modeling, in general, now exceed 2 billion US $ annually. Molecular modeling has been suggested by several professionals in food industry as a new tool for food research. Such tools can assist food scientists in problem solving, as well as save time and money. Examples of utilization of molecular modeling are the uses of high intensity sweeteners and taste receptors to predict the sweetening potential of new molecules by using molecular modeling as developed by and E.W. Taylor and S. Wilson at the University of Georgia. Such models can be used in food industry for product development and also for faster results in sensory evaluation.

Medical applications

Molecular modeling is especially helpful in medical fields, such as in development of new drugs on a nanoscale. Recent studies have shown the importance of using molecular modeling in both medical and food sciences. The molecular modeling of Epigallocatechin Gallate (EGCG), and the HIV cell was undertaken by Shearer. His report has inspired scientists in Japan who discovered the potential of green tea as an anti-HIV drug. The chemical compound that is found abundantly in the green tea called Epigallocatechin Gallate (EGCG) is reported to stop the HIV virus from binding to CD4 molecules and human Tcells.

7

Robotics in Reproductive Medicine

The word robot was defined by the Robotic Institute of America as "a machine in the form of a human being that performs the mechanical functions of a human being but lacks sensitivity...." One of the first robots developed was by Leonardo da Vinci in 1495; a mechanical armored knight that was used to amuse royalty. This was then followed by creation of the first operational robot by Joseph Marie Jacquard in 1801, in which an automated loom, controlled by punch cards, created a reproducible pattern woven into cloth. This allowed for both precision and accuracy, a recurring theme with robotic technology.

Issac Asimov elucidated the role of robotics in 1940 through short stories; however, it was his three laws of robotics that received popular acclaim. The three laws state "1) A robot may not injure a human being, or through inaction allow a human being to come to harm 2) A robot must obey the orders given it by human beings except where such orders would conflict with First Law and 3) A robot must protect its own existence as long as such protection does not conflict with the First or Second Law." These laws have pervaded into current robotic surgical systems as evidenced by their safety features.

In fields apart from medicine, the first applications of robotics were in mathematics, computers, and industry.

The first industrial robot was the "Unimate" developed by George C. Devol and Joseph F. Engelberger, which was used to extract die castings from machines and perform spot welding on automobile bodies. Currently, robotic technology is used in space and ocean exploration (taking images and collecting information), industrial tasks (welding), military and police tasks (destroying mines, collecting information, or spying), and entertainment (from toys to television).

Medicinal Applications

Robotics have recently been applied to the field of medicine, initially with rehabilitation devices and assistance for those with disabilities. Dr. David Gow created the first bionic arm in 1998 called the EMAS (Edinburgh Modular Arm System). Robotic technology can also be used to help individuals with severe disabilities perform independent activities of daily living (The Winsford feeder) or integrate them into the workplace (RAID: Robot for Assisting the Integration of the Disabled). Used in hospitals indirectly for patient care, robots such as the HelpMate operate semi- or fully autonomously as porter or couriers since 1993. Recently a new concept of "telerounding" has been introduced with a mobile robot system (InTouch Health, Santa Barbara, CA) that enables physicians to remotely interact with patients. Through a 5-ft robot, a remotely located physician can see, hear, and talk with a patient that is located on a hospital ward.

In 1985, the first surgical application of the robot was in a neurosurgical procedure. The PUMA 560 was used to orient a needle for a brain biopsy under computerised tomography (CT) guidance. However, its use was discontinued due to safety issues. Meanwhile, at the Imperial College in England in 1988, a robotic system called the PROBOT was created to aid in transurethral resection of the prostate. This was the first autonomous surgical procedure performed by a robot, where a three-dimensional model of the prostate is built, the area of resection outlined by the surgeon, the trajectories of cutting are calculated by the robot, and finally, the procedure is executed.

A few years later in 1992, International Business Machines (IBM) and associates developed a prototype for orthopedic surgery. The ROBODOC was used to assist surgeons in milling out a hole in the femur for total hip replacements. Simultaneously, the birth of robotic telepresence technology (which would allow the surgeon to operate at a distance from the operating room) was occurring at the Stanford Research Institute, National Aeronautics and Space Administration (NASA), and the Department of Defense. Originally, the prototype was created to suit the needs of the military, and the robotic arms were designed to be mounted on an armored vehicle to provide immediate operative care in the battlefield.

Soon thereafter, Intuitive Surgical acquired the prototype and commercialised the system called da Vinci, which focused on the immersive telepresence concept (where the surgeon operates at a distance from the patient, but feels as if in the operating room). At the same time, Computer Motion unveiled the first laparoscopic camera holder, AESOP (Automated Endoscopic System for Optimal Positioning). Computer Motion later created the Zeus surgical system, which is an integrated robotic system (the surgeon operates at a distance from the patient and is aware that he is at a distance).

Neurosurgery

The first clinical application for robotics was in neurosurgical stereotactic maneuvers. The original model was the PUMA 560. More recently, robotics have been used for stereotactic radiosurgery; a robot delivery system (e.g., CyberKnife) is used to manipulate a linear accelerator as it delivers closed-cranium radiation to an area identified on preoperative imaging.

Orthopedics

Total knee replacement is a common orthopedic procedure whose success is reliant on the alignment of the limb and the prosthesis. The prototype ROBODOC system uses CT guidance to mill a precise hole into the femoral cavity. A second integrated robotic system, the ACROBOT appears promising in preliminary trials. In addition, for patients with low back pain who undergo minimally invasive procedures such as nerve blocks, a robot has been developed by three different groups to assist the physician in needle placement.

Maxillofacial

A surgical robot developed in Germany uses CT scanning to perform craniofacial osteotomies with a surgical cutting saw. Preliminary studies in animals appear optimistic.

Ophthalmology

Many surgical operations on the eye require precise microsuturing skills. A robot developed by John Hopkins University is the "Steady Hand" robot for microsurgical augmentation. This robotic instrument requires the physician to actually hold and manipulate the tool with the aid of the robot. Inanimate studies evaluating the precision of suture placement have demonstrated an advantage with robotic assistance.

Urologic Applications

The urologic applications for robotic technology have ranged from transurethral prostate resections with the prototype PROBOT, to percutaneous access of the kidney for minimally invasive kidney procedures. Su et al. recently evaluated the efficacy of robotic percutaneous access of the kidney for nephrolithotomy, and found that the number of attempts and access rates were comparable to standard techniques. The role of telerobotic surgery in urology has been evaluated in a number of procedures with variable results. Sung and Gill and colleagues evaluated the role of the Zeus surgical system for pyeloplasty and ablative urologic procedures in the animal model and found increased operative times as compared to traditional laparoscopic techniques.

However, comparison of the da Vinci surgical system when used to perform robotic-assisted pyeloplasty has been shown to have comparable times to the open approach with the additional benefits of minimally invasive surgery. For radical prostatectomy in humans there appears to be comparable oncologic control with the benefits of minimal blood loss, shorter hospitalisation, quicker recovery, and early continence. Robotic surgery has also been used for laparoscopic donor nephrectomy and cystectomy, renal transplant, and recently, robotic vasovasostomy.

Gastrointestinal Surgery

The diversity of robotic cases used by gastrointestinal surgery was originally reported by Cadiere et al., and recently by the Academic Robotics Group at the Society of American Gastrointestinal Endoscopic Surgeons annual meeting. They presented their combined experience with 211 cases. These cases encompassed Heller myotomy, gastric bypass, pyloroplasty, gastrojejunostomy, esophagectomy, duodenal polyp excision, and gastric mass excision. Other procedures performed robotically include intragastric resection, distal gastrectomy, wedge resection,

Roux-en Y gastric bypass, gastric banding, biliary pancreatic diversion, rectopexy, anterior resection, and abdominoperineal resection.

Cardiac Surgery

Telepresence robotic technology was originally created for cardiac surgery. The advantages of three-dimensional visualisation and miniaturised multiarticulated instruments have allowed cardiac surgeons to perform minimally invasive endoscopic coronary artery bypass grafting and valve procedures. Currently, coronary artery bypass grafting and valvular surgery have been performed using either the Zeus or da Vinci surgical system. Results have been optimistic for both mitral valve repair in 38 patients and closure of atrial septal defects in 10 patients using the da Vinci surgical system.

Surgery in Gynecology

Surgery in obstetrics and gynecology has traditionally been taught through a laparotomy or vaginal approach. The advantages of the laparotomy approach include depth perception and tactile feedback from the resistance of tissue/organ dynamics. In addition there is an ease of intra-abdominal suturing from the 6 degrees of freedom afforded from the human wrist. Although a laparotomy is advantageous for the surgeon compared to other surgical techniques, there are disadvantages for the patient including a large abdominal incision, prolonged hospitalisation, increased postoperative analgesic requirements, and increased morbidity. This has led some surgeons to seek out minimally invasive approaches.

The first gynecologic laparoscopy was described by Ott from Petrograd who inspected the abdominal cavity using a head mirror and a abdominal wall speculum in 1901, calling the procedure "ventroscopy". However, it was the first International Symposium of Gynecologic Endoscopy in 1964 that initiated interest in laparoscopic tubal sterilisation, gamete intrafallopian tubal transfer, and other laparoscopic gynecologic procedures in the ensuing four decades. Laparoscopy offers advantages to the patient: improved cosmesis, decreased blood loss, less postoperative analgesic requirements, shorter hospitalisation time, and quicker recovery.

However, its usefulness is limited due to the steep learning curve for surgeons. Other obstacles include: limited dexterity, counterintuitive motion, two-dimensional vision, and ergonomic difficulty. Tremor amplification can also occur from the use of long rigid instruments for prolonged periods of time in a fixed position. In laparoscopic surgery, the fulcrum point created by the trocars limits the surgeon to 4 degrees of freedom, reducing dexterity. In addition, because of the fulcrum at the trocars, the movements of the surgeon's hands results in movements in the opposite direction at the working end of the laparoscope, making movements counterintuitive.

The laparoscopic surgeon must also accommodate to a twodimensional screen, which limits depth perception as compared to the three-dimensional vision afforded by open surgery. Ergonomics is also impacted by traditional minimally invasive surgery. In a survey by Society of American Gastrointestinal Endoscopic Surgeons, 8%-12% reported pain or numbness in the arms, wrists, hands, or shoulders after performing laparoscopic surgery, which has been confirmed by electromyographic data. These limitations can be overcome if the surgical procedure is facile and efficient. Simple gynecologic procedures, such as tubal sterilisation, ovarian cystectomy, and cauterisation of endometriosis, are examples of procedures that can be effectively performed through laparoscopy and have obtained popular acceptance since its first description in the 1970s.

It is the more complex, advanced laparoscopic cases, that present a challenging learning curve including microsurgical tubal reanastomosis, abdominal sacrocolpopexy, myomectomy, and radical hysterectomy. Robotic technology, more specifically telerobotic surgical systems, offers the opportunity to bridge this gap between laparotomy and laparoscopy by enabling minimally invasive surgery with three-dimensional vision, ergonomically optimal positioning, tremor filtration, and laparoscopic instruments with intra-abdominal articulation.

Robotic technology can be categorised according to their technological or functional classification. Robots can be autonomous where the preoperative plan controls the manipulator (such as with the Probot). In addition, robots can be supervisory where a computer accurately guides the surgeon (such as with the Robodoc) or teleoperated where the robot is manipulated by input devices under the surgeon's control (such as with the AESOP, da Vinci, or Zeus surgical system). Functionally, robots are classified as either active or passive. A passive robot could be used as a navigational aid or a precise positioning system (such as with the Minerva).

It is turned on for positioning and then turned off. An active robot performs the surgery by actually moving the tools (such as the AESOP, Zeus, or da Vinci system). In gynecology and its related subspecialties, there are three forms of robotic technology available to assist surgeons in performing minimally invasive procedures, all of which are examples of active robotic technology that are teleoperated.

Laparoscopic Holder

One of the first, active teleoperated robots introduced into clinical practice was the AESOP 300, approved by the Federal Drug Administration (FDA) in 1994. This robotic arm has direct control over the laparoscope and can be manipulated by the surgeon using voice control. The robot is attached to the side of the surgical table and can grasp any rigid laparoscope. The advantage of AESOP was demonstrated by Kavoussi et al. in a comparison of robotic vs. human laparoscopic camera control. In 11 patients requiring bilateral procedures, robotically controlled camera positioning was used on one side and compared to traditional hand control on the contralateral side. They found robotically controlled positioning was significantly steadier with similar operative times.

Operative times were also similar in animal studies. In gynecology, Mettler et al. used AESOP in 50 gynecologic procedures and found that the operative times using AESOP was similar to those operations using traditional hand control. Disadvantages for AESOP include the constant voice commands, which may be distracting. In addition, the voice control is slower compared to the rapid camera movements achieved by an experienced assistant. AESOP is intended to facilitate solo-surgeon laparoscopic procedures; however, the surgeon may still need an assistant for control of the fourth laparoscopic port.

Robotic Integrated Surgical System

One of two available surgical systems is Zeus developed by Computer Motion in the early 1990s. Initially, it was specifi-cally designed for cardiac operations (e.g., coronary artery bypass grafting), and later the human applicability was diversified for multiple surgical specialties such as general surgery, urology, and gynecology. The basic difference between the Zeus and other telerobotic surgical systems was that it was not developed to create an immersive intuitive interface, but an integrated robotic surgical environment, as evidenced by the original system in which instruments lacked intra-abdominal articulation and the console contained a two-dimensional viewing monitor.

The Zeus system has two subsystems "the surgeon-side" and the "patient-side." The surgeon's subsystem consists of a console that can be positioned anywhere in the operating room. The console consists of a video monitor and two handles that control the robotic arms holding the surgical instruments. The patient-side subsystem consists of three robotic arms attached to the table. Three-dimensional vision and instruments with intra-abdominal articulation capabilities were added to later versions of the surgical system. All original animal and human studies evaluating the efficacy of robotic surgery used the Zeus surgical system. However, in late 2003, Intuitive Surgical purchased Computer Motion and the Zeus surgical system. No further production of the Zeus system has been pursued.

Although gynecologists were among the first to use laparoscopic surgical techniques, the role of robotic surgery in gynecology has developed after other surgical specialties. The first use of robotics in the gynecologic literature was by Mettler et al., who used AESOP in 50 gynecologic procedures. In the subspecialty of reproductive endocrinology, the predominant procedure using robotic technology is microsurgical tubal reversals. Using the Zeus surgical system, the first procedure performed was microsurgical uterine horn anastomoses in six female pigs in 1998. This procedure capitalises on the advantages of the robotic system by providing the fine motor movements required for intracorporeal suturing, three-dimensional vision, and motion scaling to assist in microsurgery.

Falcone et al. performed the first human clinical trial using the Zeus robotic system in 1998 on 10 patients with previous tubal ligations who underwent a robotically assisted laparoscopic tubal reanastomosis. The set-up included placement of the ports in the lower quadrants bilaterally for the robotic arms and one port was placed suprapubically for introduction of suture. To perform the reanastomosis, 6-0 polygalactin was used on the mesosalpinx and 8-0 used on the fallopian tube. Chromopertubation established patency in 17 of 19 tubes reanastomosed with a pregnancy rate of 50%. They then compared their robotic reanastomoses to traditional laparoscopic reanastomosis and found that operative times were signifi-cantly longer (2 hours) with use of the Zeus robotic system, but all other outcomes were comparable.

Degueldre et al. then performed a feasibility study with the Da Vinci surgical system on 8 patients. The mean operating time was 181.5 minutes and although follow-up was limited to 4 months, 2 of the 8 patients achieved a pregnancy and 5/8 patients demonstrated at least unilateral patency. Subsequently, Dharia et al. performed a feasibility study in a fellow-ship training program using the da Vinci surgical system on 18 patients who desired reversal of tubal sterilisation and compared these to 10 patients who underwent a traditional open microsurgical reanastomosis.

Operative times were significantly greater in the patients who underwent robotic-assisted surgery, however, hospitalisation time, time to recovery, and time to return of independent activities of daily living were significantly shorter in the robotic group. Patency rates were 100% and pregnancy rates were 50%, comparable in both groups. Although limited, a preliminary cost-effective analysis demonstrates comparable cost per delivery in patients who underwent a robotic tubal reanastomosis ($92,488.00) as compared to those underwent a traditional open reanastomosis ($92,205.90). In the subspecialty of urogynecology, the role of robotics has been focused on abdominal sacrocolpopexy.

DiMarco et al. published a feasibility study on five women who underwent this procedure without complication with a mean follow-up of 4 months. In this procedure, they used conventional laparoscopy (nonrobotic) to prepare the anatomic sites, vagina, and presacral space, and used the robot to suture the mesh in place. The mean operative time was 3 hours and 42 minutes. In gynecologic oncology and benign gynecology, a pilot study on 10 female pigs who underwent hysterectomy and bilateral salpingoopherectomy using the Zeus robotic system was completed within 370 minutes with no complications and no intraoperative conversions. Subsequently, a preliminary report on 11 human patients who underwent a robotically assisted laparoscopic hysterectomy was published in 2002.

Ten women underwent a hysterectomy with salpingo-oopherectomy. Most patients required a vaginal component. One patient underwent a hysterectomy, pelvic/ para-aortic lymph node biopsies, and infracolic omentectomy. Operative times ranged from 4.5 to 10 hours (mean of 253 minutes) and estimated blood loss was 50-1,500 mL. One patient required reoperation for bleeding with subsequent transfusion. Advincula et al. reported their preliminary experience with the use of the robot for laparoscopic myomectomies. In this report of 35 patients, the mean weight of the leiomyoma was 223 ± 244 g (95% confidence interval [CI] 135-310), the mean number of leiomyomas was 1.6 (range 1-5), and the mean diameter was 7.9 ± 3.5 cm (95% CI 6.6 -9.1). The mean blood loss was 169 ± 198 mL. The mean operative time was 230 ± 83 minutes (95% CI 201-260). Five cases required between 350 and 400 minutes to complete the procedure.

There was a trend toward decreased operative times with experience. There were three conversions to laparotomy. There are no published comparative clinical trials of robotic surgery with laparoscopic myomectomy or hysterectomy. However, when we compare these robotic results to published trials without the robot there does not appear to be any advantage with these prototypes. It is possible that these robots may be more useful to the surgeon who is presently performing these procedures by laparotomy to perform them by laparoscopy rather than in the hands of the expert laparoscopic surgeon.

There are case reports of the use of the robot with other less common procedures. In a case report, the da Vinci surgical system was used without complication to perform an ovarian transposition in a patient before she received radiotherapy for a stage 1B-1 cervical cancer. Three interrupted 3-0 silk sutures were used to suture the transected utero-ovarian ligament to the psoas muscle. The role of robotic surgery has also been investigated in the pediatric and fetal population. Gutt et al. performed a bilateral gonadectomy in a 16-year-old pediatric patient with a gonadoblastoma. Using the da Vinci surgical system, the operative time was 95 minutes and no complications were reported.

An in utero sheep model demonstrated the feasibility of creation and repair of a full-thickness skin lesion (for repair of myelomeningocele) using the da Vinci surgical system. Four of the six fetal lambs survived until sacrifice. This study entertains the role of intrauterine endoscopic surgery to reduce the risks associated with fetal surgery.

Role of Robotics in Teaching/Training/Simulation

The concept of the first structured surgical training program in the United States was created by Dr. William Halsted. This concept was based on clinical service with subjective feedback from mentors (apprenticeship), and prevailed for most of the 20th century. Recently, economic constraints have focused more attention on the efficacy of surgical education. Bridges and Diamond found that education in the operating room accounted for more than 2,050 hours, which translates to $47,970 per graduating resident or $53 million per year for the surgical specialty. Other factors that affect the cost of surgical training include an increase of the per capita workload for faculty to bring in clinical dollars, decreasing time dedicated to education, decreases in Medicare graduate medical education funding, and a requirement by insurance companies for intense supervision by faculty, allowing less autonomy of housestaff.

From the viewpoint of the residents, their experience is limited by fewer working hours (80 hours/wk) and the operating room may not be a conducive environment for optimal learning. To maximise surgical training, one must identify the different components that contribute to the overall experience. Using Rasmussen's model of human behavior, surgical education can be broken down into three tiers: skill-based behavior (e.g., instrument handling/ dissection); rulebased behavior (e.g., surgical anatomy); and knowledgebased behavior (e.g., how to handle hemorrhage or unexpected situations). Currently, training models available for surgical education include animate models, cadaveric models, inanimate models (simulators and virtual reality), robotics, and the operating room. Animate models provide a realistic environment (haptic feedback, hemorrhage) that approaches the traditional operating room experience.

However, disadvantages include the cost of the animal, anatomical differences, and the ethical use of animals for training. Although cadaveric models share the same anatomy, they do not respond similarly in terms of bleeding/ haptic feedback, are costly, and require adequate facilities for maintenance. This has turned attention to inanimate models as a training tool in surgery. Inanimate models can include both simulation and virtual reality. Although not currently realistic, inanimate models can provide reinforcement or emphasis on the different levels of surgical training with immediate formative and summative assessment.

Other advantages include the ability to be used multiple times, flexibility for usage according to the provider's schedule along with good measures of reliability, feasibility, and validity. This has made surgical simulators popular in the rapidly developing area of surgical education. Another teaching tool that is emerging onto the education forefront is the use of robotics with inanimate models, animate models, cadaveric models, or patients. Robotics with intra-abdominal articulation, three-dimensional vision, and motion scaling allow for an increase in precision, which may attenuate the learning curve in minimally invasive surgical training. A survey of general surgery program directors around the country revealed that 11% of programs used robotics in their surgical practice, 14% provided resident exposure in robotic training, and just under 25% were planning to incorporate robotics in their training programs.

To evaluate whether basic laparoscopic skills could be improved with robotic technology in attending physicians, Chang et al. compared times and composite scores (involving precision, time, reliability, etc.) of traditional laparoscopic tasks, tasks using the robotic surgical system without training on the robot, and tasks after various levels of robotic training. They found that those with laparoscopic experience had equal measures of time-to-completion, but the amount of error decreased with robotic training. In surgically naïve subjects, Nio et al. found that 20 medical students, when asked to performed both basic tasks (dropping beads in receptacles or capping a hypodermic needle) and complex tasks (suturing and performing a laparoscopic cholecystectomy), demonstrated smoother and straighter movements with less grasping actions using robotics than conventional laparoscopy.

The increase in precision and accuracy was identified as an advantage by other investigators and suggests that robotic training may allow for acquisition and refinement of skill for those in training. Yohannes et al. evaluated a small group of surgeons in training and compared skills using manual laparoscopy and robot-assisted surgery. They found that beginners start off with a slower performance time, but experience a greater degree of improvement than those who have laparoscopic experience. Although both groups improve, beginners appear to learn difficult tasks easier with robotic-assisted surgery, which might make it an ideal training tool. One question that remains to be answered is whether the improvement in precision and accuracy improves performance in the operating room.

Improvement in efficiency has not been identified in all studies. However, a small study in six medical students, who used traditional laparoscopy or the da Vinci surgical system to complete basic tasks, found that the da Vinci surgical system allows laparoscopic procedures to be performed quicker and more efficiently than traditional laparoscopy. In addition, Sarle et al., when comparing 21 surgeons on eight timed drills, found that the mean time to completion of a task was faster using the da Vinci surgical system. Dakin and Gagner addressed skill outcomes of all three methods studied: traditional laparoscopy, the Zeus surgical system, or the da Vinci surgical system. Eighteen surgeons were compared based on the performance of basic tasks and fine motor tasks (intracorporeal suturing).

There was no difference in basic tasks, but there was an increased precision in fine motor tasks in the robotic groups, although time-to-task completion remained the same. Again the principal advantage was in precision. The assimilation of new technology for training can be limited by the learning curve for that instrument. The learning curve for robotics has been addressed using the da Vinci surgical system in 23 experienced laparoscopic surgeons in seven different subspecialties. Hanly et al. found that mean operative times began to decrease after the third case regardless of the procedure.

Small number of subjects studied and the lack of validation demonstrating improvement of skill in the operating room limit these preliminary studies. As the field of robotics continues to evolve, its translation to the operating room may provide a new method of training in surgical education. Currently for clinical use of the da Vinci surgical system, the FDA requires a 2-day training course to understand the set-up, maintenance, and applications of the surgical system, in addition to animate laboratory training. Further training recommendations by the company include case observations and mentored cases.

Future Robotic Surgery

The future of robotic surgery has significant potential. Currently for robotic surgical systems, areas of research include using miniaturised motors to decrease the size of the robotic tower, evaluating the potential for mounting the robotic tower to the ceiling or wall to increase access points to the patient, and real-time or non-real-time image integration into the endoscopic view on the console. Image integration would allow the surgeon to operate on the target organ and evaluate blood supply by real-time ultrasound, or identify a lesion by integrating magnetic resonance imaging (MRI)/CT images into the operative image.

From an educational standpoint, areas of active research involve adding a mentoring console to the system. This could be used for training, whereby a resident/fellow would sit at one console, the mentor at another and control over the "masters" could be toggled from mentor to student. In addition, there is research on robotic simulators, which with summative and formative assessment would have tactile feedback. Robotics in telementoring would allow for international exchange of health care ideas and new surgical techniques and provide the ability to standardise surgical training. This may also have a role in credentialing or certifying new and renewing surgeons. The potential for telepresence surgery points toward global health care. A patient with a rare disease process could be surgically treated by a world expert without travel or other barriers that may prevent health care access.

8

Application of Magnetic Techniques

Drug Discovery

The modern drug discovery process emphasises rapid data generation and analysis to identify promising new chemical entities as well as new drug targets early in the development cycle. At every step of the rapidly evolving drug discovery process, dozens of technologies and products are required. But innovations in newer technologies for genomics and proteomics are changing the face of drug discovery.

Automation has become essential in allowing researchers to meet the high through demands of today's research environment. The main thrust area where magnetic separation is applied in drug discovery is sample preparation that includes high throughput genome isolation for sequencing or PCR amplification to carry out genotyping, SNP scoring or expression profiling. The inherent benefits offered by magnetic handling includes, reduced reagent costs, elimination of labour intensive steps, easy automation and yield of high purity DNA in less amount of time compared to conventional methods.

Highthroughput DNA Isolation

Isolation of DNA is a prerequisite step for many molecular biology techniques. The separation of DNA from the complex mixtures in which they are often found is frequently necessary before other studies and procedure like sequencing, amplification, hybridisation, detection etc. The presence of large amounts of cellular or other contaminating material like proteins and RNA, in such complex mixtures often impedes many of the reactions and techniques used in molecular biology. The conventional protocol for extracting DNA involves cell lysis followed by removal of contaminating cellular components such as proteins, lipids and carbohydrates; and finally isolating DNA using a series of precipitation and centrifugation steps, which are difficult to automate. Improvement in methods for isolating DNA has been made and more recently, methods that rely on the use of solid phase have been proposed.

Adsorbents that provide fast, efficient DNA purification are important for making this procedure amenable to automation. The discovery in late 80's that silica can be used as adsorbent for DNA isolation, became the basis for most of the DNA isolation kits currently available. One of these kits involves isolation of DNA using silica coated magnetic particles. A high throughput genome isolation protocol has been developed, which is based on SPRI (Solid Phase Reversible Immobilisation) chemistry.

The SPRI protocol is based on DNA binding to the surface of carboxyl coated paramagnetic particles under the condition of high salt and PEG. The procedure has been developed and optimised for single stranded DNA isolation, such as M13 phage and double stranded plasmid DNA utilising carboxyl coated magnetic particles. The SPRI protocol has allowed the development of an automated procedure in a microplate format with a throughput of about 200, 000 DNA preparations per day, thus becoming a fastest microplate plate based DNA purification system in use on the human genome sequencing.

Magnetic Separation of Poly (A) mRNA

A number of methods have been reported for isolation of total RNA from a variety of cells or tissues. Chirgwin et al. developed a method for efficient isolation of total RNA by homogenisation in a 4 M solution of guanidium thiocyanate containing 0.1 M 2-mercaptoethanol. Homogenisation is followed by extraction of RNA by ethanol or by ultracentrifugation through cesium chloride. This method was further modified by Chomczynski and Sacchi to devise a rapid single step isolation procedure for RNA. It involves extraction of RNA using a mixture of guanidium thiocyanate and phenol-chloroform.

Many RNA isolation kits are available based on the above two protocols. All these methods isolate RNA on the basis of its biochemical properties. In contrast, biomagnetic separation of mRNA is based on specific complementary hybridisation between poly A sequence of isolated mRNA and oligo $(dT)_{25}$ sequence covalently linked to the surface of paramagnetic particles.

In this method oligo $(dT)_{25}$ coated magnetic beads are added to crude cell or tissue lysate. During incubation poly (A) mRNA from the lysate are caught onto the surface of oligo $(dT)_{25}$ coated magnetic beads. The beads/mRNA complex is then washed magnetically. The mRNA thus isolated is either eluted or directly applied for many downstream applications which includes cDNA library construction, Substractive hybridisation, Northern hybridisation, RT-PCR and in vitro translation.

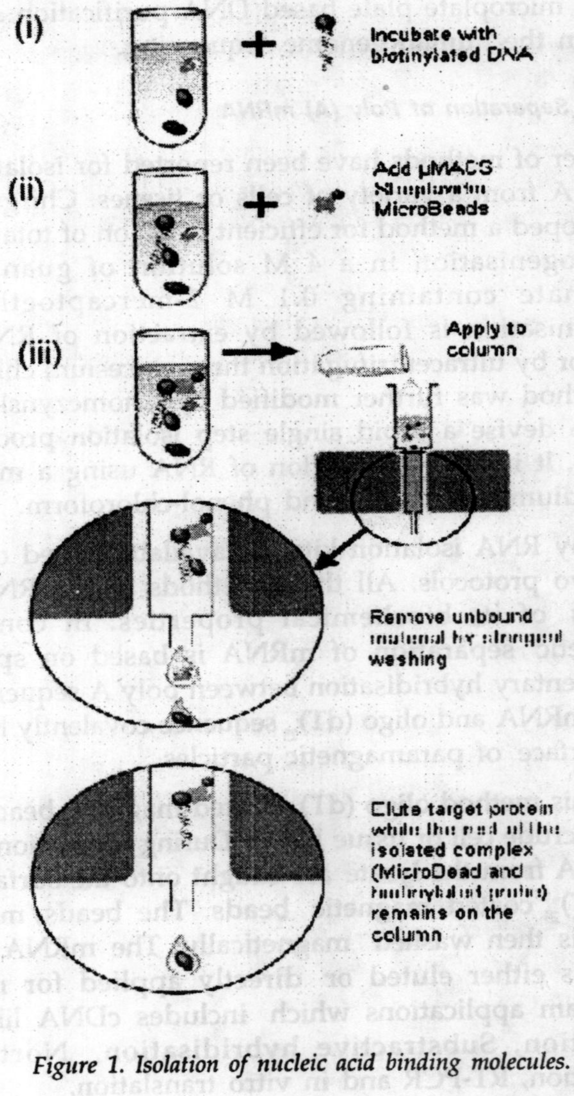

Figure 1. Isolation of nucleic acid binding molecules.

Isolation of Nucleic Acid Binding Biomolecules

The isolation of specific molecules based on its interaction with complementary binding partner is emerging as important technologies in many field of research. The isolation and characterisation of specific transcripts or proteins can be employed to monitor the progression of disease. Several kits are available in the market that works on the principle of magnetic labeling and direct isolation of biotinylated molecules such as DNA, RNA or proteins onto streptavidin coated magnetic beads (μMACS streptavidin Microbeads from Miltenyi Biotec, Germany). These biotinylated molecules can then be used for indirect isolation of non-biotinylated target molecules that may interact with them.

The procedure involves complex formation between the biotinylated probe (DNA, RNA or proteins) and the target molecule (i.e. interacting biomolecules DNA, RNA or protein). Based on the interaction of biotin-streptavidin, the probe-target complex is then separated from rest of the component by addition of streptavidin coated magnetic beads. The complex is magnetically isolated and washed to remove non-specifically bound molecules.

The non-biotinylated target molecules can be either eluted off from the complex with high purity, whereas the magnetically labeled biotinylated probe remains bound to the column (Fig. 1). This technique has potential for rapid and efficient screening of transcriptional and translational regulatory proteins. Alam and co-workers have also used target DNA-conjugated magnetic beads for rapid screening of DNA-binding peptide ligands from solid phase combinatorial library.

Proteomics Applications

Magnetocapture Protein Interaction Assays

Magnetic particles are now increasingly used as carriers

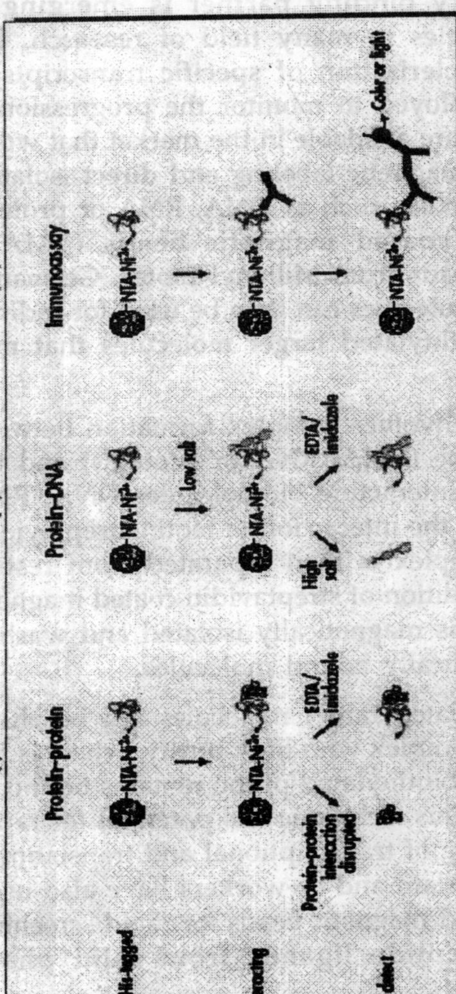

Figure 2. Magnetocapture protein interaction assays.

for binding proteins, enzymes and drugs. Studies have shown that proteins and enzymes can be bound covalently to naked magnetic particles in the presence of carbodiimide. Such immobilisation procedures for proteins, enzymes or drugs will have a major impact in various areas of medicine and biotechnology. The immobilised biomolecules can be used directly for a bioassay or as affinity ligands to capture or modify target molecules or cells. On this basis, Ni-NTA (nitriloacetic acid) tagged magnetic agarose beads have been used for versatile magnetocapture assays using 6xHis-tagged proteins.

The procedure involves use of metal chelating nitriloacetic acid (NTA) groups covalently bound to the surface of agarose beads, which contain strong magnetic particles. The beads are precharged with nickel, which is ready to capture 6xHis-tagged proteins for sensitive interaction assays or microscale purification of 6xHis tagged proteins (Fig. 2). Thus, this technique bridge the gap between purification-scale procedures using Ni-NTA metal chelate affinity chromatography resins and microplate-based assays. Other procedures for magnetic affinity separations of various proteins have been summarised recently.

Biomedical Applications

Drug Targeting

Magnetically guided drug targeting has been attempted in order to increase the efficacy and reduce the unpleasant side effects associated with chemotherapy. This method of drug delivery involves immobilisation of drug or radionuclide in biocompatible magnetic nano- or microspheres. This method of delivery makes chemotherapy more effective by increasing the drug concentration at the tumour site, while limiting the systemic drug concentration. FeRx Inc. of San Diego have designed Magnetic Targeted Carriers termed MTCs for site-specific targeting, tissue retention, and sustained release of drugs. These MTCs are composed of elemental iron particles and activated carbon. MTCs (1-2 μm in size) can adsorb and desorb pharmaceutical agents such as doxorubicin (DOX).

Vialed MTCs are mixed with an anticancer drug already in solution; the mixture is then introduced into the catheter. The drug delivery using MTCs includes insertion of catheter into an arterial feed to the tumor, followed by application of powerful magnetic field to cause the MTC-DOX to extravasate through the capillary bed into the targeted tissue (Fig. 3). The field is left in place for another 15 minutes; after removal of the magnet, the particles remain trapped in the tumor, where the drug is then released.

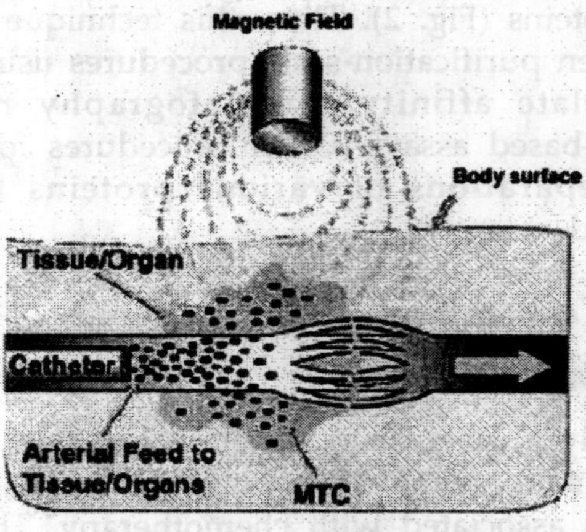

Figure 3. Drug targeting using MTCs.

FeRx has already initiated phase III clinical trial for MTCDOX formulation. Like wise, Hafeli and co-workers have carried out preclinical studies in collaboration with

FeRx for treatment of liver and brain tumors using MTCs labeled with β-emitters such as Rhenium-188 and Yttrium-90. Although the main research concern of MTC-drug therapy is cancer treatment, the technology is not limited to this, other drugs that are deliverable in this manner include antibiotics, thrombolytics, anti-inflammatories, peptides and steroids.

Magnetoliposomes

Magnetoliposomes are magnetic derivatives of liposomes and can be prepared by entrapment of ferrofluids within the core of liposomes. Several groups have investigated the uses of magnetoliposomes for site specific targeting, cell-sorting and as magnetic resonance contrast enhancing agents. Babincova et al. have tried to use magnetoliposomes with encapsulated doxorubicin (DOX) for site specific chemotherapy in response to externally applied AC magnetic field. The result of this study showed that magnetoliposomes can be specifically heated to 42°C in a few minutes and during this, the encapsulated doxorubicin is massively released.

Another in vivo studies for site specific targeting using magnetoliposomes incorporated with adriamycin (ADR), showed that administration of magnetic ADR liposomes under an applied external magnetic force produced approximately 4-fold higher maximum ADR concentration in the tumor than did administration of ADR solution alone. These results suggested that systemic chemotherapy could effectively control the primary tumor without significant side effects, due to the targeting of magnetic ADR liposomes. Magnetotactic bacteria belonging mainly to Magnetospirillum species produces ferromagnetic crystalline particles consisting of magnetite (Fe_3O_4) or greigite (Fe_3S_4). These are called bacterial magnetic particles (BMPs).

The BMPs are small in size (50-100 nm) and disperse very well because they are covered with a stable lipid membrane. Magnetoliposomes containing cis-diamminedichloroplatinum (II) (CDDP) prepared using BMPs have also been evaluated for targeting and controlled release of drugs at tumor site. It has been observed from such studies that the capture volume of the magnetoliposomes with BMPs was 1.7 times higher than the ones with artificial magnetic particles. Also, when a rotating magnetic field was applied to these magnetoliposomes, the contents of them were released over 2 hours, suggesting that magnetoliposomes containing CDDP shows antitumor activity and that the controlled release of drugs from magnetoliposomes is possible.

Magnetic Fluid

Magnetic fluids have been investigated as potential hyperthermia causing agents due to their high specific absorption rate (SAR). Hyperthermia is a promising approach for cancer treatment that uses AC magnetic fields to heat target areas (cancer tissue) containing magnetic fluids. To study the biological effects of AC magnetic field excited ferrofluids, both in-vitro and in-vivo studies have been carried out in cancer cell lines and spontaneously induced tumors in animal models. The results of these studies have shown that magnetic fluid hyperthermia is able to reduce the viability of cancer cells, thereby indicating the potential of this therapy.

There also exist the possibility of combination therapy, which would include hyperthermia treatment followed by chemotherapy or gene therapy. The approach involves use of magnetic nano- or microspheres or magnetoliposomes containing a drug to cause hyperthermia using the standard procedure, followed by the release of encapsulated drug that will act on the injured cells. It is anticipated that the combined treatment might be very efficient for treating solid tumor.

Cell Sorting Applications

Immunomagnetic Cell Isolation and Separation

Immunomagnetic cell separation methods have become increasingly popular among the cell biologist. The general approach involves use of paramagnetic particles coated with antibodies against the target specific cell surface molecules. The method employs two schemes for isolating the target cell, direct method that involves coupling of affinity ligand (antibody) onto the magnetic particles, which are then directly added to the sample containing the target cells. During the incubation the magnetic particles binds the target cells that can then be recovered using a magnet. While in case of indirect method, a free affinity ligand (an appropriate antibody) is first added to the cell suspension.

After incubation excess unbound affinity ligand is removed by washing, and the antibody-target cell complex is then captured by magnetic particles bearing an affinity ligand (secondary antibody) with the affinity for the primary label. Both positive and negative selection can be performed with immunomagnetic separation. In case of positive selection target cellular subsets are magnetically labeled and subsequently separated, while negative selection involves target purification by removing all other contaminating cells.

Using this technology large number of cell types has been isolated to date. Some of the examples where immunomagnetic separation technology has been applied includes, detection and removal of circulating tumor cells from peripheral blood, isolation of antigen specific CD 8+ T cells, selective separation of CD34+ cells, isolation and enrichment of retinal ganglion cells (RGCs) for culturing etc. Immunomagnetic separation is used extensively for detection of pathogenic bacteria in clinical, food and environmental samples. In addition to whole cell isolation, even cell organelles can be selectively separated using magnetic particles. The lysosome fraction was isolated from the amoeba Dictyostelium discoideum after feeding with dextran-based nanoparticles and subsequent homogenisation. Likewise a variety of cell types can be isolated using automated systems and different surface coated magnetic particles that are available from Miltenyi Biotec, Germany, or StemCell Technologies, USA.

Immunomagnetic Assays

Immunoassays are becoming an important tool in clinical diagnostics and in fundamental investigations, because of its sensitivity, specificity and general applicability. Magnetic particles have been applied increasingly for various immunoassays that include fluoroimmunoassays, enzyme immunoassays or radioimmunoassays. Magnetic particles bound with primary or secondary antibodies are used for separation and quantification of antigens.

Use of magnetically bound antibodies elimirates the centrifugation step thus reducing assay time and simplifying operation thereby increasing the efficiency and accuracy of the assay. Matsunaga and co-workers have developed a novel fluoroimmunoassay method, where a fluorescein isothiocyanate (FITC) conjugated monoclonal anti- Escherichia coli antibody was immobilised onto bacterial magnetic particles (BMPs) for detection and removal of E. coli. The same group has also developed a chemiluminescence enzyme immunoassay with BMPs using IgG as a model antigen. Likewise a magnetic bead based ELISA has been developed for detection of Staphylococcus species.

9

Regulation of Nutraceuticals and Pharmaceuticals

Nutraceuticals and pharmaceuticals are often mentioned in one breath. The reason is that these products have physiological activity in human and animals, and all are related to health in its broadest sense. Irrespective of their exact definition, nutraceuticals and pharmaceuticals can be distinguished at best on the basis of their direct aim: is the product a food, or is the product a medicine. Nutraceuticals aim at maintenance of the health situation of healthy individuals.

Nutraceuticals add health promotion to the regular function of a foodstuff being a source of building material and energy. Nutraceuticals are available over the counter, e.g. in a supermarket, and can be consumed as part of the daily diet. Pharmaceuticals are produced to cure human or animal individuals, for mitigation or for diagnostics and are generally available from a doctor or veterinary surgeon. Some pharmaceutical products, especially those for mitigation and self-diagnostic purposes, are freely available.

The aims and controlled availability make the position in daily life of pharmaceuticals clear, in contrast to the position of nutraceuticals. Pharmaceuticals are medicinal preparations. Their effects and side-effects in human and animals must be known exactly from medical and veterinary trials. Nutraceuticals have primarily to be proven safe-to-eat; of many nutraceuticals, the beneficial activity as such has not been demonstrated unambiguously.

Nutraceuticals and Functional Food Crops

There is a lot of terminology around nutraceuticals which may raise considerable confusion. Synonymo us to, or belonging to the same field of terminology of 'nutraceuticals' are 'functional foods', 'designer foods', 'positive nutrition', 'foods with dietary supplements', 'foods with functional ingredients', 'health food', 'dietary food', 'fuctional food ingredient', etc. Nutraceuticals are colloquially understood as a product that can be a single well-defined food compounds with health promoting characteristics, but also as complex foods with such beneficial characteristics.

Neutraceuticals may be briefly and meaningless defined as 'nutritionally or medicinally enhanced foods', whereas other authors give more extended descriptions. Anyway, the term 'nutraceutical' originates from DeFilice, founder of the Foundation for Innovation in Medicine. He defined nutraceutical as: 'food, or part of food, that provide medical or health benefits, including the prevention and treatment of disease'. The first part of this definition relates to a complete (complex) food produc t or a single compound. The latter part makes a nutraceutical similar to a pharmaceutical. On the other hand, Schaafsma defines 'nutraceutical' (written as nutriceutical) in a narrower sense as the specific purified or extracted food supplement, tha t is also available separately (encapsulated) from a drug store or a reform shop (over the counter).

Examples are vitamins, minerals, unsaturated fatty acids, milk proteins, etc. Further he defines foods supplemented with specific nutraceuticals to meet a particular objective as 'designer foods', whereas 'functional foods' include all health promoting foods with 'designer foods' among them. Different definitions distinguish between nutraceuticals, as food products that function as (and therefore substitute) pharmaceuticals, and functional foods, as foods with human health enhancing qualities. The problem is that these terminologies and definitions blur the border between foods and medicins. Such situation should be prevented.

A food can be made a functional food by using different approaches:

— to eliminate a component known to cause deleterious effects to the consumer (e.g. an allergenic protein),

— to increase the concentration of a natural component in food,

— to add a component which is not normally present in most foods, but for which beneficial effects have been demonstrated (e.g. non-vitamin anti-oxidants, or fructans),

— to replace a component, usually a macronutrient, the intake of which is usually excessive (e.g. fats) by a component which has beneficial effects (e.g. oligosaccharides), and

— to improve the bioavailability of, or to modify, food components for which beneficial effects have been demonstrated Role of genetic modification

For several of the approaches given above, genetic modification can be used to improve a food crop directly into a functional food crop:

— antisense technique for elimination of health-impairing proteins;

— overexpression to increase the concentration of specific functional ingredients;

— introduction of a new metabolic pathway or side chain;

— blocking of a metabolic pathway.

Compounds with health promoting properties that can be considered for inclusion into a nutraceutical or a functional food are, for example, flavonoids, carotenoids, unsaturated fatty acids, oligosaccharides, fibres, and minerals. Flavonoids may inhibit the development of cancer and arteriosclerosis. Carotenoids can scavange toxic oxygen radicals and function as provitamins. Multiple unsaturated fatty acids may prevent heart and vascular diseases. Oligosaccharides and fibres can bind toxic compounds and may serve as food for, and this way improve the quantity and quality of, the intestinal flora.

Oligosaccharides and fibres are poorly digestible and are therefore helpful in keeping the dietary energy low. In addition, health effects have been attributed to minerals, including the minerals calcium (bone development), selenium (as an anti-oxidant) and iron (to prevent anaemia). Increasingly, transgenic research in plants is directed to engineer the relevant metabolic pathways.

Functional foods and nutraceuticals are becoming more and more popular with consumers who desire to influence their own health and well-being through their diet. On the other side, there are the food producing industries that experience new market niches for their products when they are upgraded to a functional food. The scientific basis for the application of functional foods is much less clear than that of (recombinant) pharmaceuticals and (genetically modified) medicinal (food) plants. At present, for most existing functional foods adequate scientific bases for their health benefits are lacking.

Nutraceuticals and functional foods both have, by definition, positive effects on consumers' health. The issue of health claims that are attached to these products is therefore important. In short, there is no law in The Netherlands on the use of health claims, but voluntary procedures have been installed by three institutions that are active in this field. Important in this respect is the weight of scientific evidence that can be provided in support of the health claim.

Pharmaceuticals and Medicinal Plants

Pharmaceuticals are single, well-defined medicinal and therapeutic compounds that are administered with the purpose to cure, mitigate, or diagnose disease. Similar compounds for veterinary application are also categorised to pharmaceuticals. Medicinal plants are plants that produce pharmaceuticals naturally. Pharmaceuticals can be produced chemically, or biologically by micro-organisms, animal cell cultures, and plants. In these biological systems, the pharmaceutical can be a natural product, or can result from genetic modification. Plants are one of the new hosts that can serve for the production of recombinant pharmaceuticals. Biologically produced pharmaceuticals are often indicated as biopharmaceuticals.

Recombinant Pharmaceuticals in Crop Plants

Currently, the production of recombinant pharmaceuticals in crop plants includes especially proteinaceous compounds such as hormones, antibodies, plasma proteins, ensymes and vaccines. The production of oral vaccines with transgenic crop plants offers the advantage that the crop needs not be purified to be consumed as a vaccine, such as raw tomatoes and bananas. Such specialty crops enable to combine experiences from the production of pharmaceuticals in microbial and animal cell production systems with experiences on the genetics and agricultural characteristics of the crop involved, including the typical biological and production characteristics.

As for other genetically modified crops (such as those producing nutraceuticals), human and environmental safety during production must be assessed before such plants can be approved as specialty crops in agriculture (greenhouse or open field). Such assessment will show which concerns have practical relevance, and enables to set up strategies to deal with these concerns. Risk assessments on genetically modified crops in general, including pharmaceutical-producing plants, always need a case-by-case approach using sound scientific and clinical judgement.

For the largescale agricultural production of recombinant pharmaceuticals, the choice of plant species, geographic location of test plots and production fields, the choice for production under contained greenhouse conditions, and monitoring protocols are essential to prevent adverse consequences. With a qualified harvest of plant material (e.g. seeds), the challenge is to convert an agronomical commodity into a pharmaceutical, that is essentially identical to the same product from any other production system. This requires every step in the production process to be described as good manufacturing practice (GMP).

Comparing recombinant nutraceuticals and pharmaceuticals produced in plants, the nutraceuticals are often products from the primary or secondary metabolism of the plant cell whereas pharmaceuticals may also consist of proteinaceous compounds directly resulting from an introduced single transge ne. Metabolic pathway engineering in plants will in future enable to produce exotic and rare compounds at increased scales.

Medicinal Plants

Medicinal plants are known from a long history. Many plant species have been named according to some medical application such as 'Hepaticae' (liver mosses) as old medicine against liver diseases. And a common species name of many plants is 'officinalis' which refers to medical application. About one-third of medicines used today are derived from plants, with aspirin, morphine, artemisin and taxol as wellknown examples. Three possibilities are considered here for the novel product derived from genetically modified plants:

1. The product is a "functional food / feed / crop plant"; the product is used for food purposes (food, food ingredient, a dietary food) or animal feed purposes (feed, feed ingredient).
2. The product is a "nutraceutical"; the product is used for food purposes (food ingredient, food additive, food supplement) or animal feed purposes (feed ingredient, feed additive).
3. The product is a pharmaceutical and is used as a medicinal preparation.

For the various shapes that the novel product from a genetically modified plant may take within any of these possibilities, different regulations apply both in the EU and the USA.

Uses of Food

EU

A food or food ingredient derived from a genetically modified organism is considered "novel" in the EU. Other foods considered "novel" in the EU are foods that have not been previously consumed within the EU, derived from micro-organisms, significantly altered, or processed in novel ways. Novel foods should be evaluated for market approval under EU Novel Food Regulation 258/97. Please note that viable genetically modified products need an approval for marketing under EU Directive 90/220 prior to an approval for food applications under the Novel Food Regulation.

There is, however, an additional short-cut route of "notification". In the case of "notification", the applicant supplies evidence for the "substantial equivalence" of a product derived from genetically modified organisms to a conventional / commercial counterpart. Notifications that have led to market approvals include those for processed oil from genetically modified canola and products, such as starch and oil, from genetically modified maize kernels.

For the "authorisation" procedure, decision trees will provide guidance on the data required for compiling the dossier. In all cases, information should be provided on the toxicological and microbiological safety of the novel food, in addition to its nutritive value and its estimated intake. Substantial equivalence is the central principle in the assessment of the safety of a genetically modified crop, based upon which it is decided which safety tests will be required. The submitted crop or its products are substantially equivalent to a conventional counterpart if they are phenotypically and compositionally comparable.

In that case, no further safety testing would be required under the present policy. If, however, the crop or product is substantially equivalent except for one or a few compounds, the toxicological significance of the altered compounds should be assessed. Crops or foods/ feeds that are not substantially equivalent at all should be submitted to a full safety test. Food additives are defined as substances that do not occur normally in consumed foods, excluding, however, flavourings and, in general, nutrients (minerals, vitamins). If, by coincidence, the nutraceutical possesses sensory properties, it may be used as a food flavour. Flavourings are used to add odour or taste to foodstuffs. Several substances are allowed as such under EU Directive 88/388, including substances isolated from vegetable tissue by physical means.

Specific permissible food uses of sweeteners, colourants, and other additives are described by EU Directives 94/35, 94/36, and 95/2. For additives and flavourings to be approved, no distinction is made in EU legislation for substances derived from genetically modified organisms. Foods and genetically modified foods containing detectable transgenic proteins or - DNA should be labelled under EU Regulation 49/2000. The same applies to foodstuffs containing additives or flavourings derived from genetically modified organisms under EU Regulation 50/2000, provided the additives or flavourings have been released on the market after April 2000.

USA

Genetic modification is not considered different from other technologies for crop improvement in the USA. American legislation is therefore oriented to the safety evaluation of the altered characteristics of the genetically modified food. Newly expressed gene products may be regarded "food additives" or "pesticide products" and should therefore be evaluated for their food safety. Food additives are evaluated by the FDA and pesticide residues by the Environmental Protection Agency (EPA). American food additives should always be evaluated for their safety by the FDA, while other foods and food ingredients should be "generally recognised as safe" (GRAS) under the Federal Food Drugs and Cosmetics Act (FFDCA).

The determination of GRAS status by the FDA is not mandatory. Moreover, a food substance can be declared "GRAS" by its manufacturer ("GRAS self-determination"). In practice, however, the market will only accept GRAS determinations that are endorsed by recognised institutions in order to avoid legal liability for the use of food substances that may turn out to be unsafe. These recognised institutions include the FDA itself, but also "extramural" associations such as the Federation of American Societies for Experimental Biology (FASEB) and the Flavor and Extract Manufacturers' Association (FEMA). The procedure for food additive petitions is more tightly regulated than for GRAS status.

The FDA's Redbook lists the toxicological tests required for safety testing of food additives and - ingredients. Newly expressed gene products such as viral coat proteins, viral replicases, herbicide degrading enzymes and herbicide tolerant enzymes are regarded "plant pesticides". These gene products are therefore evaluated by the EPA, which determines the pesticide levels that are tolerable in raw agricultural commodities. The FDA will, in turn, accept the occurrence of these pesticides in food products, provided the pesticide levels in these products do not exceed the tolerable levels determined by the EPA.

Transgenic proteins that have thus been tolerated by the EPA in commodities are, however, generally exempted from an upper threshold level of tolerance based on the absence of harm from the estimated human exposure to these pesticides, which are present at low levels in plants. The FDA recently made a proposal for new legislation, under which the notification of genetically modified food / feed crops by their manufacturers 120 days before market release will become mandatory. In addition, rules are developed for those producers that wish to label genetically modified foods / feeds voluntarily.

Special Uses

EU

Foods for infants, sportsmen, and hospitalised patients (e.g. enteral nutrition) are among the "foods with particular nutritional uses" as defined by the EU. Such foods should be labelled as "dietetic" or "dietary". Special requirements for the levels of vitamins and minerals in infant formulae have been formulated in amendments to EU Directive 89/398, which regulates these types of food.

USA

Similar to the EU's provisions for foods for particular nutritional purposes, the FDA has formulated requirements for the manufacturing-, composition-, and labelling- of "foods for special dietary use", including infant formulae and weight- loss products. In addition, "medical foods" for oral consumption or enteral nutrition have been defined by American law as foods that provide nutrients to people other than healthy, such as those with diseases or "distinctive nutritional requirements". Medical foods are exempt from the regulatory requirements for food labelling, including health claims (FDA, 2000c).

Food Supplements

EU

Food supplements are defined by the EU as concentrated forms of nutrients, which serve to supplement the dietary nutrient intake, such as vitamins, minerals, amino acids, fatty acids, fibre, and herbal- and plant extracts. An EU Directive on Food Supplements will likely enter into force in the near future, as a proposed version was recently adopted by the European Commission. For the time being, the directive gives a limited account of vitamins and minerals permitted to be sold as supplements.

Requirements for other ingredients, such as plant and herbal extracts, are still to be defined. For those supplements that are not covered by EU Directives, national legislation of EU member states should apply. Purified vitamins and minerals can also be added to foods rendering these foods "fortified". Member states' regulations on food fortification have not been harmonised yet at an EU level.

Netherlands

Food supplements are considered "health products", i.e. products that look like pharmaceuticals or that are linked to he alth- functions, but that are no medicines. Similar to conventional foods, health products are regulated by the Commodity Act. Food supplements containing vitamins are more specifically regulated by the Commodity Act Rule waiving Vitamin Preparations ("Warenwetregeling Vrijstelling Vitaminepreparaten"). Food fortification is regulated by the Commodity Act Decision on the Addition of Micro-Nutrients to Foods ("Warenwetbesluit Toevoeging Micro-voedingsstoffen aan Levensmiddelen"). The addition of all micro- nutrients is permitted, except for iodine, fluorine, and amino acids. Three categories of food fortification with micro-nutrients are considered:

— *Restoration*: compensating for nutrient losses during food manufacture.

— *Substitution*: addition of nutrients of a product to a substitute for that product.

— *Enrichment*: addition of nutrients to products that do not contain them. Limits are posed to the added amounts of micro-nutrients, based on their recommended daily intakes. Fortified foods are allowed to bear claims pertaining to their nutrient contents.

USA

American food supplements may consist of vitamins, minerals, herbs and dietary components as tablets, fluid or food. Food supplements are not allowed to be marketed as foods. Supplements are less stringently regulated as they fall under the Dietary Supplement Health and Education Act. Their constituents need not be GRAS, but the manufacturer should be able to prove their safety. In case the supplement contains a novel ingredient, the FDA should receive a notification and data substantiating the novel ingredient's safety 75 days prior to its market release American legislation on food fortification is similar to that of the Netherlands.

In addition, efforts are made by the EU and the US to harmonise legal criteria for food supplements containing vitamins and minerals, in order to ensure marketability of these products in both regions. These criteria will include, among others, Good Manufacturing Practice (GMP).

Health Claims

EU

There are three types of claims, which are: nutritional-, health- and ethical claims. Nutritional claims pertain to the composition of the foodstuff, while health claims concern the potential health benefits of consumption. Ethical claims say something about the working conditions under which the food has been manufactured. Two directives are most tightly linked with food labelling, 2000/13 (labelling and advertising of food) and 90/496 (labelling of nutritional value). These directives are, however, not comprehensive on health claims. EU Directive 2000/13, for example, does only define those claims that are not permitted as follows: "The labelling and methods used must not ... attribute to any foodstuff the property of preventing, treating or curing a human disease, or refer to such properties".

Hence genuine medicinal claims are not allowed on foods. This may change, however, as a recent resolution of the Euro-parliament attached a high priority to the inclusion of all labelling issues into one directive. EU Directive 90/496 regulates nutritional claims on labels. Such claims relate to the nutritional value of the foodstuff, such as the content of vitamins. It does not, however, specify rules for the evidence needed in support of such claims. Labelling of dietary foods is regulated by EU Directive 1999/21, which extends EU Directive 89/398 on foods for particular nutritional purposes. Health claims for these foods are subject to pre- market clearance.

The legislation on claims in individual EU member states may be more extensive than the rules set out by the EU directives. Scrutiny of health claims may be performed either pre- or post-clearance of the food product. The European Commission recently proposed to harmonise legislation on:

1) *Nutritional claims*: A draft list has been provided, featuring a number of permissible nutritional claims.

2) *Functional claims*: i.e. health claims that relate to the positive effect of a nutrient on a bodily function.

Netherlands

Dutch legislation on nutritional claims follows EU Directive 90/496 and pertains only to the nutritional value of foodstuffs. In the Netherlands, a distinction is made between "health products" and medicinal products. "Health products" bear the shape of a pharmaceutical, such as tablets and tinctures, but they are distinguished from medicines by the attached claims that relate to health without making them medicines. It should be noted that food supplements too fall under the Dutch definition of health products. Health claims for health products can be scrutinised voluntarily. For this scrutiny, three institutions have each developed their specific code of practice. These institutions are:

1) Advertising Code Foundation in which the media and consumers participate
2) Verification Council in which branch organisations (producers, media, ma rketeers, drug retailers) participate.
3) Nutrition Centre, which is independent and disseminates public information on nutrition and food safety. In all three procedures, relevant scientific data must support the proposed health claim.

A Code of Practice has been developed by the Nutrition Centre, which specifies the criteria that should be fulfilled by the scientific evidence in support of a claimed health benefit of food and drink products. Adherence to the Code is on a voluntary base. Criteria for the advertisement of "health products", i.e. products with either a pharmaceutical appearance or attributed health-function, have been formulated by the Verification Council. One important criteria in the Verification Council's evaluations is that no pharmaceutical- like properties may be ascribed to the health product.

Complaints about these advertisements can be assessed by the Advertising Code Committee, which verifies their compliance with the Dutch Advertising Code. Previous evaluations by the Verification Council of health claims submitted to the Advertising Code Committee will be taken into account. Claims pertaining to vitamin preparations, as a category of food supplements, are specifically regulated by the Commodity Act Rule waiving Vitamin Preparations ("Warenwetregeling Vrijstelling Vitaminepreparaten"). These claims have to be supported by scientific evidence, that the manufacturer or trader must be able to deliver at request.

In addition, fortification of products with micro-nutrients and the claims pertaining to fortified products are regulated. Fortification is allowed if its purpose is to compensate for micro-nutrient losses during processing or to create a substitute for another product containing the micronutrient, i.e. vitamin- fortified margarine vs. milk butter. Several micro- nutrients are prohibited (e.g. iodine) or allowed only in small amounts (e.g. vitamins A-D) due to their narrow safety margins. Sponsorship of non-profit health associations provides for an additional opportunity to attract consumers' attention to health-promoting food products. The Dutch Heart Foundation allows companies to display the foundation's logo on food products exclusively for fund-raising acitivities.

Several conditions must be fulfilled, however, to obtain permission to use the logo. The food product in question should i) be a food of preference for the Nutrition Centre, or ii) bear a heart-related health claim that has been positively evaluated according to the Nutrition Centre's Code of Practice. In addition, the manufacturing company's image should be positive and compatible with a healthy lifestyle (no tobacco companies allowed, for example). The logo should be displayed such that its conjunction with fund-raising is obvious to the consumer and that it's not mistaken for a quality mark. As example, Unilever's phytosterol-enriched margarine spread Becel Pro Activ was allowed to bear the Heart Foundation's logo during several months while the foundation was receiving royalties from the sales of Becel Pro Activ. In a similar vein, the Dutch Digestive Diseases Foundation offers sponsoring companies the opportunity to seal certain food products with the foundation's logo against a fixed fee under the following terms:

— The product should be compatible with the foundation's message; for example, it should be high in fibre and/or moisture.

— Claims should be scientifically supported.

— The foundation logo symbolises the company's support for the foundation and should not be mistaken for a quality mark.

USA

In the United States, a distinction is made for permissible health claims on foods and on food supplements. Please note that the American definition of a health claim is narrower than ours and limits itself to "substance-disease" claims. Three types of claims are discerned for both foods and food supplements:

1) "Substance-disease" claims, which are also referred to as "health claims" in the USA. These claims pertain to the beneficial effect of a food substance on human health without becoming medicinal (i.e. to cure, mitigate, or diagnose disease). A limited number of acknowledged health claims are permitted on foods, such as prevention of cancer by fruit consumption. Fewer acknowledged health claims are allowed on food supplements than on foods. Novel substance-disease claims can be cleared for labelling through three regulatory pathways:

   — Claims may be submitted to the FDA, which will evaluate the scientific background supporting these claims. The FDA requires that "significant scientific agreement" exists for such claims to be cleared.

— Claims on foods (not on dietary supplements) may be based on authoritative statements from scientific bodies, such as the National Academy of Sciences (NAS) and federal institutions such as the National Institutes of Health. These claims should reflect consensus within these scientific bodies and be based on review of the scientific evidence. The following approved claim, for example, was based on a statement by NAS: "Diets rich in whole grain foods and other plant foods and low in total fat, saturated fat, and cholesterol, may help reduce the risk of heart disease and certain cancers".

— "Qualified" claims are allowed on food supplements (not on foods) as a result of a court decision on a recent lawsuit between a supplement manufacturer and the FDA (the "Pearson vs. Shalala" case). The FDA was required by the court to permit claims that did not match the FDA's standards for "significant scientific agreement", provided that qualifying language was used that would render these claims not misleading. The qualifying claims that have since then been approved by the FDA contain either qualifying remarks (rendering claims rather lengthy) or disclaimers.

2) "Nutrient content" claims on foods and food supplements (such as "high in calcium"), which pertain to the level of a nutrient for which a recommended daily intake or a daily reference value has been established. Such claims have to be submitted to the FDA for evaluation. In addition, nutrient-content claims may be based on authoritative statements, similar to substance-disease claims.

3) "Structure function" claims on foods and food supplements describing the relation between a nutrient

and body structures, physiological functions, nutrition deficiencies, and common affections that are associated with "passages of life". Claims related to severe diseases are, however, precluded. Structurefunction claims are not liable to the pre- market scrutiny and should be notified to the FDA within 30 days after market introduction. The burden of proof for these claims lies on the side of the FDA, which has to react to non permissible claims. A disclaimer should be added that, among others, the FDA has not evaluated the claim.

In addition, requirements for ingredient labelling of dietary supplements are less rigorous than for foods and allow for the omission of non-relevant ingredients and for the declaration of special nutrients. In addition to explicit health claims, several American producers of foodstuffs, food supplements, and pharmaceuticals are allowed to seal the logos of non-profit health organisations to their products and advertisements in return for sponsorship. An example is the labelling of Florida fruit products with health logos sponsored by the US State of Florida's Department of Citrus.

Florida orange juice labels are allowed to display logos of the American Cancer Society (reduced risk of cancer) and the March of Dimes (reduced risk of birth defects), while Florida grapefruit is allowed to bear the American Heart Association's logo (reduced risk of heart disease). It should be noted that the American Cancer Society will not accept new sponsors for this purpose. The American Heart Association, on the other hand, has a food certification programme in place where food manufacturers can acquire permission to label their products with the "heart-check mark". Eligible food products must be low in fat, cholesterol, and salt, but contain enough nutrients.

Fruit juices, fortified beverages, and lean beef are among the food products that have thus been approved. Logos of other American non-profit health organisations not mentioned here have also been linked to products and advertisements. It should be borne in mind, though, that American consumer laws and Federal Trade Commission standards apply to the use of logos as described above. Both the logoproviding non-profit health organisation and the sponsoring food manufacturer have the responsibility to ensure that the logo display will not mislead the consumer. If, for example, the health organisation and the manufacturer have entered an exclusive agreement, this should be stated clearly. In addition, if the health organisation does not specifically endorse a labelled product, this should be stated too.

Feed Uses

EU

A list of materials, including plant products, that are allowed to be marketed as ingredients of compound animal feeds in the EU, is provided within EU Directive 91/357. Additional clearance under EU Directive 90/220 is required for animal feeds and feed ingredients derived from field-tested and commercially released genetically modified crops, as long as the crop is considered a viable organism. A specific EU Novel Feed Regulation, however, is anticipated to replace the animal feed evaluation under 90/220 in the EU in 2001. Contrary to the 90/220 directive, this novel regulation also covers the use of non viable GM products and requires post market monitoring and tracing methods for the novel feed.

Netherlands

Same as the EU for viable products (cultivation and/or import). For the import of non viable animal feed products, a voluntary notification can be made to the Dutch Ministry of Agriculture, Nature Management, and Fisheries. The outcome is not legally binding and does not apply to other EU Member States.

USA

American regulations on animal feed differ from their European counterparts. Feed ingredients that are allowed to be included in animal feeds are listed in the "Official Publication" of the Association of Animal Feed Control Officials (AAFCO), a voluntary organisation that advises authorities' officials. Each State of the USA may have its own feed regulations, additional to those of the Food and Drug Administration (FDA). "Nutraceuticals" as feed ingredients have apparently not been regulated yet, given the discussion inside AAFCO on whether or not nutraceuticals should be registered as feed ingredients and, consequently, evaluated for their safety.

Feed Additives

EU

Feed additives from genetically modified organisms that are similar to any on the market are not specifically regulated and are regarded as safe. The presence of transgenic DNA containing antibiotic resistance genes is, however, not allowed. Proposals have been amended by the Europarliament to install mandatory labelling of feed additives that contain- or consist of- genetically modified organisms. This would imply that purified additives lacking transgenic components would not fall under this labelling requirement.

USA

Feed additives are considered animal medicines in the USA, hence they should be approved as a new animal medicine. The marketing of a feed additive that is generally recognised as safe (GRAS) by the FDA will, however, only require an expert's opinion, eventually supported by literature data.

Medicinal Uses

EU

Approval for the production of medicines by genetically modified organisms can only be granted by a "centralised" EU procedure as described by EU Directive 2309/93. A comprehensive dossier should support the application for a medicine derived from genetically modified organisms. Guidance for applicants on the content of dossiers is provided by the European Agency for the Evaluation of Medicinal Products (EMEA). Three types of data must be collected for the marketing application dossier: quality-, preclinical-, and clinical- data. Quality includes medicine composition and stability, while special requirements have been formulated for vegetable products.

Preclinical data comprise in vitro data, such as mutagenicity, and animal test data, such as pharmacokinetics and toxicity; the latter includes single dose- and repeated dose- tests, teratogenicity, and carcinogenicity. Specific guidance is provided for the preclinical testing of biotechnology-derived medicines. Clinical trials are human trials in which pharmacokinetics, efficacy and toxicity are assessed. The order in which data are collected is not random, i.e. clinical trials are only permitted after safety and efficacy in animals have been assessed by preclinical trials. Human medicinal products are classified into two categories by EU Directive 92/26:

1. Prescription medicines, which can only be sold through pharmacies.

2. Non prescription ("over the counter", OTC) medicines, i.e. medicines that can be obtained with- or without- a physician's prescription.

Several criteria are considered before a medicine is granted the OTC status, including the safety of the medicine, the ability of the patient to self-diagnose and to self- medicate with little or no professional instructions. Advertising is only possible for OTC medicines, not for prescription medicines. After their market release, medicines should be monitored for side effects. The "market authorisation holder", together with health professionals and government officials should report side effects to the Member States and EMEA. Serious adverse effects have to be reported within 15 days, whereas other side effects are reported periodically. No special additional requirements have been formulated for the monitoring of biotechnology-derived medicines.

Netherlands

Advertisements for OTC medicines are under scrutiny of the Verification Council (Keuringsraad KOAG/KAG) in the Netherlands. The Dutch Medicine Evaluation Bureau recently advised the Dutch Minister of Health Welfare and Sports to allow the general sales (i.e. outside pharmacies and drug stores) of certain OTC medicines, such as vitamins and minerals. In the Netherlands, pharmacovigilance is carried out by health professionals, medical specialist organisations, the marketing authorisation holder, and official bodies. For human medicines, these official bodies are the Medicine Evaluation Bureau and the Health Inspectorate.

USA

Under American legislation, the same stages in medicine testing prior to marketing can be discerned as in EU legislation: quality, pre-clinical, and clinical. In the clinical stage, the medicine is tested as an "investigational new drug (IND)". All relevant details of the preclinical tests and the IND-tests should be submitted in the "new drug application (NDA)". If the NDA is approved by the reviewers, the new medicine is allowed to be marketed. A short-cut route can be followed if the medicine for whic h a marketing approval is sought is a "generic drug", i.e. a medicine that has already been approved and that lacks patent protection. This route is named "abbreviated new drug application (ANDA)".

Apparently, USA legislation does not exempt biotechnology-derived medicines from this possibility, unlike EU legislation. For an ANDA, the "innovator" medicine should be proven to be bio-equivalent to the approved medicine, i.e. it should display the same uptake from the dosage. Bio-equivalence studies should involve 24-36 human volunteers, in which the bloodstream uptake of the medicine is measured. These studies can substitute for preclinical- and IND-studies. ANDAs for "biologics" (e.g. proteins), however, are still a matter of discussion in the USA, as reported recently.

Similar to the EU, the FDA discerns "prescription (Rx)" medicines and "over the counter (OTC)" medicines. OTC medicines are not classified into active compounds but into their indication or purpose. At present, "OTC drug monographs" are compiled, which stipulate composition and quality of approved OTC medicines. If a new medicine's composition concurs with that described in an OTC monograph, a simplified OTC medicine application can be filed instead of an NDA. Botanical products that have been marketed as food or food supplements and that are "generally recognised as safe (GRAS)" by the FDA can also be sold in the USA as OTC medicines.

Similar to the EU, pharmaceuticals must be monitored post- market for adverse effects. The FDA has programmes in place for reporting adverse reactions, viz. the Adverse Event Reporting System for manufacturers, packers and distributors, and MEDWATCH for health professionals.

Regulatory authorities of the EU, USA, and Japan have joined forces in the International Conference on Harmonisation (ICH) to harmonise the requirements for medicine applications. The ultimate goal is the mutual acceptance of medicine application data that have been compiled in accordance with ICH guidelines. Consensus has already been reached on requirements for preclinical and clinical data. Moreover, ICH guidelines have been prepared for the production of biotechnological medicines from genetically modified cell cultures.

Veterinary Medicines

EU

The requirements for dossier data for the registration of veterinary pharmaceuticals are quite similar to those for human pharmaceuticals. Human toxicity is also relevant to veterinary medicines with regard to consumer safety. It should be noted that for veterinary medicines a maximum residue limit in food animals must be defined according to EU Regulation 2377/90. In addition, a withdrawal period has to be determined during which the medicinal product should not be administered to food animals prior to slaughter. Similar to human medicines, veterinary medicines can be either "prescription" or "OTC", as EU Directive 90/676 defines which medicines can only be administered on prescription. OTC status of veterinary medicines has not been harmonised, though, in the EU.

Netherlands

Veterinary medicines that can only be dispended through the pharmacist or the veterinarian have been defined by the "canalisation" rules under the Dutch Law on Veterinary Medicines. After the market release of veterinary medicines, adverse reactions should be reported to the Centre for Veterinary Pharmacovigilance.

USA

The American legislation on veterinary pharmaceuticals by and large resembles that of human pharmaceuticals. Hence American legislation discerns "investigational new animal drugs (INADs)", "new animal drugs (NADs)", and "abbreviated new animal drug applications (ANADAs)". Guidance documents on the various issues surrounding veterinary medicines, i.e. specific uses and clinical testing, have been published by the FDA's Center for Veterinary Medicine. Similar to the EU, a distinction is made between prescription ("Rx")- and OTC veterinary medicines.

A veterinary medicine may be changed, however, from prescription- to OTC- status if long-term experience warrants its safe use. Once a medicine has been approved for commercial release, post- market reporting to the FDA's Center for Veterinary Medicine of adverse reactions of veterinary medicines is mandatory for medicine manufacturers and voluntary for animal owners and veterinarians.

Pharmacoeconomics

One aspect that is currently receiving much attention of professionals involved in the human medicine approval process is that of health economics, also known as pharmacoeconomics. Pharmacoeconomic studies on pharmaceuticals take various economic and social aspects of the novel medicine into account, such as its cost effectiveness and its effect on the "quality of life". Cost effectiveness include the pricing of the medicine compared to other medicines, but also savings made in other sections of healthcare, such as hospitalisation of patients, by use of the novel medicine. Cost-effectiveness considerations are nowadays pivotal for a medicine to get incorporated, for example, into reimbursement schemes of health insurance providers.

There is no common EU guidance on the issue of pharmacoeconomics that we are currently aware of. Guidelines have been issued for pharmacoeconomic data to be supplied with the medicine application in The Netherlands. It may be worth noting that dietary preparations are reimbursed by Dutch healthcare insurance companies under certain conditions, e.g. if they alleviate severe diseases, such as cow's milk allergy in children till 2-3 years of age. The Dutch organisation for healthcare insurance companies is currently reviewing this policy of reimbursing dietary preparations. In the USA, pharmacoeconomic data provided by medicine companies to committees and other bodies that decide on eligibility of the medicine for therapy or reimbursement are considered "promotional" data.

Biosafety Considerations

Case-by-case

Concerning recombinant production, the specific characteristics of the crop and the recombinant compounds, the culture requirements, and the (national and international) guidelines and regulations make each production system an individual case. Biosafety assessments therefore needs to be approached on a 'case-by-case' basis. This will be the starting point.

Crop Type

Any crop type, i.e. any food, feed and industrial crop, has the potential to be genetically modified to produce pharmaceuticals or nutraceuticals. Relevant considerations before choosing a specific crop for transgenic medical or health purposes is the familiarity with the agronomic characteristics of the crop, specific methods of crop protection, weed/pest management and possibilities for the removal of agrochemical residues. Major attention should be given to the required containment conditions and the biological containment characteristics of the crop.

The biological characteristics of the plant in the field, such as dispersal of pollen, seeds and other diaspores, and its hibernation capacity or the invasiveness are relevant in this respect. Outcrossing to adjacent fields with the same crop cultured for traditional purposes must be prevented. The same holds for outcrossing to wild or weedy crossable relatives. Also possible new effects, due to the presence and characteristics in the plant of the specific health compound, to organisms that accidentally feed on it or live in its phyllosphere or rhizosphere should be considered. Several production systems, especially in the case of pharmaceutical production, will require defined greenhouse facilities.

The type of the functional food, nutraceutical, or pharmaceutical in the plant and its aim needs to be considered. In the case of pharmaceuticals it may be an antibiotic, a hormone, an enzyme, a vaccine or an immunological protein. Active components in functiona l food crops can involve flavanoids, carotenoids, multiple-unsaturated fatty acids, oligosaccharides, fibres but also increased mineral contents and vitamin level. Each type of compound, when present in unnatural quantities, or in a new host plant, may require its own safety approach. In the same vein, the "transgene-centred approach to biosafety" described in scientific literature may serve as an auxiliary tool for the biosafety assessment of genetically modified plants. This approach enables to answer concrete questions about ecological and toxicological consequences of the presence of gene xxx or gene product XXX and their metabolites in a genetically modified plant.

Attention should also be given to the safety of the farmers. Different requirements for handling during culture and harvest may be needed for a crop in its new destination as a medicinal plant or functional food crop. Each crop produces residues that can play a role in further food chains, as feed, or as manure. The presence and the concentration of a (new) medical or health compound in these residues should be analysed, and its possible effects should be considered. In the case of residues from a food crop used for the production of a pharmaceutical, it might be used for further purposes e.g. as feed or as green manure. In such case, the pharmaceutical should be removed or inactivated completely.

Greenhouse

The demands put on a greenhouse suited for culturing genetically modified plants for the production of nutraceuticals, functional foods and pharmaceuticals depend on both the legislation related to the GMO status of the produced plants and on specific production requirements claimed by the company that processes the harvested transgenic plant material. For many manufacturing processes these production requirements are laid down in EU Directives describing principles and guidelines of "good manufacturing practices" (GMP). Unfortunately, no GMP directives exist about premises and equipment for transgenic plants that produce pharmaceuticals or nutraceuticals. Production of pharmaceuticals involves legislation related to medicines.

Production of nutraceuticals involves legislation related to foods, including novel food. Currently, the EMEA (The European Agency for the Evaluation of Medicinal Products) Working Party on Herbal Medicinal Products agreed to a proposal for guidance on good agricultural and collection practice in relation to the Good Agricultural Practice (GAP) guide for Active Pharmaceutical Ingredients. The GAP guidelines apply to the production of all plant materials used in the food, feed, medicinal, flavoring and perfume industries. It applies to all methods of production including organic production in accordance with the European regulations. Producers, traders and processors of medicinal and aroma tic plants should comply with the GAP guidelines by a Way Bill (batch documentation) and demand that their partners also meet these requirements.

Combination of the requirements laid down in the GAP guidelines and to the Directives related to the transgenic status of the plant can be used to outline the conditions under which genetically modified plants can be grown in greenhouses for medicinal production. The latter, however, poses a problem: no directives are made that deal with the production of transgenic plants in greenhouses. EU Directive 90/220 is about the deliberate release into the environment of genetically modified organisms and was recently amended by EU Directive 2001/18, which has to be implemented in Member State legislation in 2002. EU Directive 90/219 is about the contained use of genetically modified micro-organisms. Production of pharmaceuticals or nutraceuticals in greenhouses is not about deliberate release in the environment. In contrast, it can have a high level of containment (especially under GAP conditions).

On the other hand, contained use of micro-organisms involves techniques and risks that are very different from contained growth of genetically modified plants. To fill up this gap in legislation, EU member states have made legislation based on EU Directives 90/219 and 90/220 for the production of genetically modified plants. EU Directive 90/220 consists of four parts.

The first part contains the general provisions: member states must take appropriate measures to avoid adverse effects on human health and the environment which may arise from deliberate release of GMOs in the environment. The second part contains the provisions for the deliberate release of GMOs into the environment for research and development purposes. The third part contains provisions for market release of GMOderived products and the fourth part contains additional procedural considerations.

Legislation in relation to production facilities in the USA is essentially comparable to that in Europe. In The Netherlands, the COGEM (commission on genetic modification) deals with applications for use of genetically modified organisms. The COGEM evaluates applications via a case by case scenario. The criteria that are used are given in the "Rule on genetically modified organisms and the pertinent COGEM guidelines". COGEM describes the conditions for greenhouses, climate rooms etc. for production of transgenic plants.

The greenhouse facilities needed for transgenic plants that produce pharmaceuticals or nutraceuticals are similar. First, the implications of the GAP guidelines for genetically modified plant production will be discussed. The consequences of the principles and guidelines for GAP for greenhouse construction will be addressed in relation to the different phases of plant production and plant processing. Whilst the recommendations set out below should be considered to be generally applicable, individual products may present particular quality control issues. Thus, the production and control of each product must be given careful individual consideration taking fully into account any special features.

The occurrence of not species/variety- identical plants and parts of plants has to be controlled in the course of the entire production process. This can easily be performed in greenhouses (better than in the field). No special demands have to be made regarding greenhouse constructions.

Cultivation

Depending on the mode of cultivation of transgenic plants for production of nutraceuticals or pharmaceuticals, growers should follow certain Standard Operating Procedures for cultivation. These are not described in any guidelines or directives but have to be elaborated by the growers themselves. In general, care should be taken to avoid environmental disturbances. Greenhouses can be designed in such a way that the probability for such disturbances is at an acceptable level. For instance, the use of gauze in front of windows and the connection of the entrance of the greenhouse compartments to a corridor (atrium) which can only be entered via an air- lock ensures that no insects can enter and that no pollen can exit the greenhouse.

Care has to be taken that no sludge, soil contaminated with heavy metals, residues of plant protection products or any other not naturally occurring chemicals can enter the greenhouse. Irrigation water should be as free as possible from contaminants, such as faeces, heavy metals, pesticides, herbicides and toxicologically hazardous substances. Pesticide and herbicide application should be avoided as far as possible. The use of pesticides has to be documented. It is obligatory that the buyer of the batch be informed of the brand, quality and date of pesticide use in a written form. In greenhouses the use of soil type, fertilisers and pesticides can easily be optimised (especially compared to transgenic plants that grow in the field). In fact, nowadays in the case of regular crop production in greenhouses these demands are already met.

Harvest

In the course of the harvest, care should be taken to ensure that no toxic weeds can mix with the harvested crop. Damaged and perished plant parts must be promptly eliminated. The harvested crop must be protected from pests, mice/rodents and domestic animals. Pest control measures should be documented. All these requirements have no extra effects on the demands on greenhouses facilities.

Primary processing includes steps of processing such as washing, freezing, distilling, drying etc.. All these processes must conform to European and national regulations. Depending on the production process, primary processing (such as drying) may take place in special compartments of the greenhouse. In many cases specialised companies will conduct primary processing. Requirements for the location for primary processing are comparable to those mentioned under "cultivation".

Packaging

The harvested product should be promptly packaged according to the demands of the pharmaceutical company that processes the plant material. This can be performed in special compartments of the greenhouse.

Storage and transport should be according to the demands of the pharmaceutical company that processes the plant material. This can be performed in special compartments of the greenhouse. Equipment should be easy to clean in order to eliminate the risk of contamination. All machinery (preferably non-wooden) should be mounted in an easily accessible way and must be well serviced and regularly cleaned. Fertiliser and pesticide application machinery must be regularly calibrated.

Personnel should receive adequate education before performing tasks that require this knowledge. All processing procedures should fully conform with both EU guidelines on food hygiene (among others EU Directive 93/43) and the general principles for food hygiene of Codex Alimentarius as well as EU Directive 91/356 on good manufacturing practice for pharmaceuticals. Personnel should be required to have a high degree of personal hygiene (toilets must have hand washing facilities etc.). Persons suffering from known infectious diseases transmittable via food, including diarrhoea, or being transmitters of such diseases, must be suspended from areas where plant processing takes place or wear appropriate protecting clothing, until their complete recuperation.

Crop Plants Producing rFSH

A recombinant form of follicle stimulating hormone (rFSH) serves as model for a pharmaceutical protein produced by a hypothetical genetically modified crop plant. Follicle stimulating hormone (FSH) is a dimeric glycoprotein that induces ovulation. Together with luteinising hormone it constitutes "gonadotropin". Pharmaceutical preparations may contain either "chorionic gonadotropin" or "menopausal gonadotropin". Chorionic gonadotropin is produced by trophoblast cells and is present in high amounts in serum and urine of females after 7-12 weeks of pregnancy. In addition, serum and urine of postmenopausal females contain elevated gonadotropins, which may serve as source of menopausal gonadotropin.

FSH, LH, and gonadotropin have found application for infertility treatment in human females involved in in vitro fertilisation therapy, while in animals such as cows it is used as inducer of superovulation. In addition, some preparations are prescribed to treat male infertility as well, since FSH stimulates testosterone production. FSH cannot be applied within the crop as food or animal feed because FSH is orally inactive, which is most likely caused by intestinal degradation of FSH.

In most approved medicine applications, FSH is injected subcutaneously, intramuscularly, or intravenously. FSH for pharmaceutical purposes has been isolated from female body fluids and - parts, such as urine, serum, and pituary glands. The advantages of FSH produced through recombinant DNA technology, however, are that i) the FSH preparation will be free of antagonistic luteinising hormone activity and ii) that pure FSH can be combined with other hormones in any desired ratio. The commercial recombinant FSH products are produced by genetically modified Chinese hamster ovary (CHO) cell lines. An additional advantage of the envisioned production of rFSH in crop plants rather than in CHO cells is the avoidance of the risk of transmission of mammalian pathogens (e.g. viruses, prions) through CHO cell culture, which eliminates the costs inherent to the efforts to control this risk under good manufacturing practices.

An additional application for rFSH may be in pregnancy testing as reference material for in vitro diagnosis of human chorionic gonadotropin (hCG) or its subunits in female urine, plasma, or serum. Kits for hCG-diagnosis are available "over the counter" for clinical- and physician's office- laboratories in the USA. To acquire marketing permission for such kits, the specificity and sensitivity of the test should be demonstrated in clinical trials. In the EU, pregnancy tests fall under the definition of an in vitro diagnostic medical device and are thus regulated by EU Directive 98/79.

For pregnancy tests, a distinction will be made between devices for self-testing (by users at home) or laboratory use. In the case of a self- testing device, a "notified body" (e.g. an independent certifying institution) must evaluate the technical data and the quality assurance programme for production. Laboratory tests for pregnancy are exempt, however, from evaluation by a notified body. Both self- testing- and laboratory- pregnancy tests should acquire a CE marking for compliance with European standards.

In The Netherlands, production conditions for a genetically modified crop can either be in-doors ("contained use") or out-doors ("environmental release"). The transgene product rFSH will possess endocrine activity, hence open- field cultivation of the novel crop may prove unfeasible unless its environmental hasard, e.g. for wildlife, can be negated. It may be worth noting here that FSH is orally inactive, hence consumption of the crop is unlikely to cause pharmacological effects. In the case of "contained use", measures should be taken to avoid environmental escape of viable genetically modified plant material during handling of the harvested crop, else this will be regarded "environmental release". The timesavings and lesser dossier requirements for "contained use" should be waged against the lower costs of field cultivation, i.e. "environmental release".

Application for Medicinal Use

FSH and rFSH have already been approved for use as human- and veterinary medicine. In the EU, full dossiers should be compiled for all biotechnology-derived medicinal products, including pre-clinical and clinical trial data, regardless if they may be "bio-equivalent" to previously approved products. Our hypothetical plant-derived rFSH preparation will therefore require full testing.

A "short-cut" route for bio-equivalent biotechnology-derived "biologics" appears possible in the USA, but FDA's policy on this issue is still evolving. In the most optimistic setting, less extensive testing will be required for the hypothetical "bioequivalent" rFSH in the USA than in the EU.

rFSH has so far been produced by genetically modified cultures of Chinese hamster ovary (CHO) cells. The fact that FSH is a glycoprotein may complicate the demonstration of "bio-equivalence", as glycosylation in plants may differ from that in, for example, hamster cells.

It may be worth mentioning here that the difference in chemical composition between a "bio-equivalent" urinary go nadotropin and a reference medicine already approved for medicinal use has been disputed over in a lawsuit between two pharmaceutical companies in the USA. The court finally ruled in favour of the "bio equivalent" gonadotropin. The disputed difference in protein glycosylation between the FSH molecules was described as "microheterogeneity".

The heterogeneity between crop-plant-derived rFSH and CHO-cell-derived rFSH may, however, be larger than the microheterogeneity observed between the urinary FSH molecules. It remains therefore uncertain if this lawsuit has set a precedent for our hypothetical rFSH preparation, if this were to be clinically equivalent to the currently approved CHO-cell-derived rFSH preparations. To avoid conflict with the protection afforded by patents and exclusivity terms to some FSH preparations already in the market, care should be taken in choosing the appropriate formulation and uses of the hypothetical rFSH.

With regard to the choice of production conditions, some reference is provided by an EU guideline on the manufacturing through genetically modified cell lines of cytokines, which are comparable to FSH in that they are physiologically active glycoproteins. It is recommended herein to pay attention to the post-translational modification of the cytokines under the manufacturing conditions, and to avoid contamination with viruses and DNA. Each batch should also be tested for activity in animals.

10

Convergence of Modern Biology and Medicine

Nanobiotechnology is defined as a field that applies the nanoscale principles and techniques to understand and transform biosystems (living or non-living) and which uses biological principles and materials to create new devices and systems integrated from the nanoscale. The integration of nanotechnology with biotechnology, as well as with infotechnology and cognitive science, is expected to accelerate in the next decade. The convergence of nanoscale science with modern biology and medicine is a trend that should be reflected in science policy decisions.

Nanotechnology provides the tools and technology platforms for the investigation and transformation of biological systems, and biology offers inspiration models and bio-assembled components to nanotechnology (Figure 1). Investigative methods of nanotechnology have made inroads into uncovering fundamental biological processes, including self-assembly, cellular processes, and systems biology (such as neural systems). Key advances have been made in the ability to make measurements at the subcellular level and in understanding the cell as a highly organised, self-repairing, self-replicating, information-rich molecular machine.

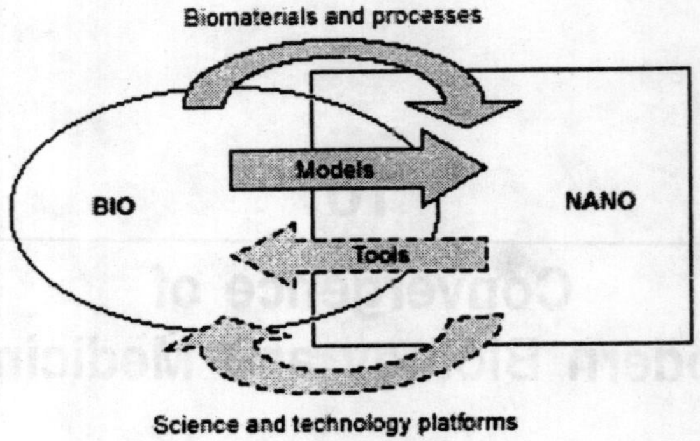

Figure 1. Key interactions between fields of biology and nanotechnology.

Single-molecule measurements are shedding light on the dynamics and mechanistic properties of molecular biomachines, both in vivo and in vitro, allowing the direct investigation of molecular motors, enzyme reactions, protein dynamics, DNA transcription and cell signaling. It has also been possible to measure the chemical composition within a single cell in vivo. Measurements of the intermolecular mechanics of a single protein molecule, polymer molecule or 'soft' nanoparticle have been performed with atomic force microscopy (AFM). Nanoscale instrumentation has also allowedmeasurements of smallRNAs (also called 'nanoRNAs' or short stretches of RNAranging in length between 21 and 28 nucleotides) and their significant effect on gene expression.

The discovery of these small RNAs was selected as the Science 'Breakthrough of the Year' in 2002. Furthermore, a biological force microscope, which is capable of quantitatively measuring interfacial and adhesion forces between living bacteria and mineral surfaces in situ, has been developed. Another contribution of nanotechnology was the development of a nanoscale system that displays protein unfolding events as visible color changes. This system will greatly enhance our ability to visualise structural changes of proteins in complex synthetic and living systems. Rectified Brownian motion has been used to explain several chemomechanical energy conversions typical to intracellular processes, as well as the kinesin motion along mictrotubules.

Spatial and temporal interaction among cells, including intracellular forces have also been measured. AFM has been used to obtain the intermolecular binding strength between a pair of molecules in physiological solutions, providing the quantitative evidence of their cohesive function. Flow and forces around cells have been quantitatively determined and the mechanics of biomolecules are now better understood. It is accepted that cell architecture and macro behavior is determined by small-scale intercellular interactions.

Fluorescent semiconductor nanoparticles, or quantum dots, have been developed for use in imaging and have been employed as markers for biological processes. These nanoparticles offer several advantages over fluorescent dye molecules, as they photobleach much more slowly and their emission wavelength can be finely tuned. Key challenges for the further development of quantum dots relate to their encapsulation with a biocompatible layer and the need to avoid nonspecific adsorption. Understanding systems biology, such as the neural system, is enabled by measurements made at the level of developing interneuronal synapse circuits and their ~20 nm diameter synoptic vesicles.

Advances in nanotechnology have allowed such measurements to be made. Nanoscience investigative tools have helped us to understand self-organisation, supramolecular chemistry and assembly dynamics, and have furthered our knowledge of the self-assembly of nanoscopic, mesoscopic and even macroscopic components in living systems. However, current understanding of the biosystem building blocks at the nanoscale is far from complete. As an illustration, one cannot explain the anomalous behavior of water with small impurities that is an essential component in a living system. Also, key brain functions have yet to be discovered. Emerging areas in which nanotechnology is set to play a role include the realistic molecular modeling of soft matter, obtaining non-ensembleaveraged information at the nanoscale, understanding energy supply to and energy conversion in cells (e. g. using photons or lasers), and the study of tissue regeneration mechanisms.

Nanotechnology offers new solutions for the transformation of biosystems and provides a broad technological platform for applications in several areas: bioprocessing in industry; molecular medicine (e. g. for the detection and treatment of illnesses, body part replacement and regenerative medicine, nanoscale surgery, synthesis and targeted delivery of drugs); investigating the health effect of nanostructures in the environment (e. g. pollution by nanoparticles and eco-toxicology); improving food and agricultural systems (e. g. enhancing agricultural output, new food products, food conservation); and improving human performance (e. g. enhancing sensorial capacity, connecting brain and mind, integrating neural systems with nanoelectronics and nanostructured materials).

Several examples of applications of nanotechnology are given in Box 1. Nanotechnology has also enabled the development of biochips and has a role in green manufacturing (e. g. biocompatibility and biocomplexity aspects). Other applications have included the design of sensors for astronauts and soldiers, biofluidics (for handling DNA and other molecules), in vitro fertilisation of live stock, nanofiltration, bioprocessing 'by design' and traceability of genetically modified food. Exploratory areas for nanotechnology will include research into the condition and/or repair of the brain and other areas for regaining cognition. It might also find application in designing pharmaceuticals as a function of patient genotypes and in applying chemicals to stimulate production as a function of plant genotypes.

The synthesis of more effective and biodegradable chemicals for agriculture and the production of implantable detectors could be aided by nanotechnology. Employing this technology it should also be possible to develop methods that use saliva instead of blood for the detection of illnesses or that can perform complete blood testing within one hour. Broader issues include economic molecular medicine, sustainable agriculture, conservation of biocomplexity, and enabling emerging technologies.

Biomimetics is a frequently used term to describe the use of concepts and principles from nature and their application to creating new materials, devices and systems. Emulating the concepts and principles of biology has led to the controlled self-assembly of biopolymeric materials, of arrays of nanoparticles, of devices for use in nanoelectronics, and of macromolecular crystals. Inorganic systems have also been assembled at the nanoscale using bioagents. Self-assembly and self-organisation have inspired ideas for 'bottom-up' manufacturing and other bioengineering methods at the nanoscale.

Bottom-up manufacturing based on selfassembly is the most promising assembly method in the long term; however, the approach has not yet been successful in assembling more than one scale for useful systems. Biological models cannot be copied directly, because biosystems work on a much smaller time scale than usually necessary in industry and need water when most industrial processes take place in other media. Creating larger molecular structures, structure replication, neuromorphic engineering, mimicking photosynthesis, and the development of bioreceptors and biomarkers, are further examples of creating artificial nanoscale devices and semibiological hybrids using biological paradigms.

Nature provides biologically assembled materials and biology's molecular toolbox has been used to develop biohybrid processes and products. Biostructures and bioprocessing offer a relatively large number of opportunities with industrial and medical relevance. These include the use of organic-inorganic hybrid materials and the creation of nanostructures as building blocks (so-called 'molecular Lego'), which can be used to make, for example, nanodevices for biosensing or coatings. Exploratory research in this area will focus on creating architectures in multiscale systems and on the creation of sensors with improved characteristics.

The United States have initiated a multidisciplinary strategy for the development of science and engineering fundaments through the National Nanotechnology Initiative (NNI). Japan and Europe have broad programmes and their current plans look ahead four to five years. More than 35 countries have developed programmes in nanotechnology since 2000, illustrating the importance of this field of research. Research on biosystems has received increased support in 2002, as compared with previous years, in the US, UK, Germany, Switzerland and Japan. Other significant investments in nanotechnology research programmes with contributions to nanobiosystems have been made in the EC, Australia, Taiwan, Canada, Finland, Italy, Israel, Singapore and Sweden.

Relatively large investments in nanotechnology with a small biosystems component have also been made in South Korea and China. Worldwide government funding of nanotechnology has increased about five times since 1997, exceeding $2 billion in 2002. Differences among countries are observed in the research domain they are aiming for, the level of programme integration into various industrial sectors, and in the time scale of their R&D targets. In the US, the Federal Government investment in nanotechnology R&D increased from $116 million in fiscal year (FY) 1997 to about $697 million in FY 2002.

The establishment of the NNI was announced on January 21st 2000, by then President Clinton, and has received support from President Bush's Administration. It includes fundamental research on nanobiosystems and three bio-related research 'grand challenges' on novel materials, health and nanobiodevices. The NNI currently coordinates the nanoscale research interests and programmes of 17 organisations through the Nanoscale Science and Engineering Technology (NSET) subcommittee of the National Science and Technology Council (NSTC). NSET sets the long-term vision, plans, budgets and programmes and carries out evaluations to ensure a successful initiative. The subcommittee is composed of representatives from each participating agency, the Office of Science and Technology Policy, and the Office of Management and Budget.

The largest investments in FY 2002 were made by the National Science Foundation (NSF) ($204 million), the Department of Defense (DOD) (about $200 million), the Department of Energy (DOE) ($89 million), the National Institutes of Health (NIH) ($59 million), the National Institute of Standards Technology ($77 million) and the National Aeronautics and Space Administration (NASA) ($35 million). Other agencies with a nanotechnology investment in the same FY are the Departments of Agriculture (USDA), Justice (DOJ) and Transportation (DOT) and the Environmental Protection Agency (EPA).

States, universities and industry have provided funding for infrastructure, education and small business development in an increasing trend.

The 10-year vision for nanotechnology published in Nanotechnology Research Directions included the study of nanobiosystems in both fundamental research and in a section dedicated to nanotechnology applications to biology, medicine and health. The interdisciplinary NNI definition of nanotechnology has promoted the integration of physical, chemical, biological, material science and engineering aspects. Nanotechnology research represented about 0. 6% in Federal R&D funding in the US in FY 2002. The NSF is the largest investor in nanoscale science and engineering with $221 million in 2003, which represents about 5% of its overall R&D budget. Of the total NNI investment in 2002, about 12% is explicitly dedicated to nanobiosystems in two ways.

First, the implementation plan of the NNI has a focus on fundamental research on nanobiosystems (about 8% of the total NNI budget in FY 2002). Second, grand challenges on 'health issues and 'bionanodevices' cover about 4%of theNNI budget. Additional investments have been made for development of the respective infrastructure in various centers including the Cornell Nanotechnology Center and Nanoscale Science and Engineering center at Rice University, in education and societal implications studies. In FY 2002, the nanobiodevices grand challenge was refocused on nanobiodetection and protection. The NNI was evaluated by the National Research Council (NRC) and its findings published in June 2002.

One of the recommendations was to expand research at the interface of nanoscale technology with biology, biotechnology and life sciences. Such plans to extend nanobiosystems research are under way at the DOE, NIH, NSF and USDA. An NSF-Department of Commerce (DOC) report recommends a focus on improving physical and mental human performance. NSF, NASA and the DOD have included aspects of converging technologies and improving human performance in their programme solicitations. The Defense Advanced Research Projects Agency (DARPA) has a programme on 'Engineered biomolecular nanodevices/systems'.

The NNI has inspired or stimulated nanotechnology programmes worldwide. As a reflection of its interdisciplinary characteristics, nanotechnology is supported by dedicated programmes that cross a range of fields or in materials, life sciences, environmental issues and electronics in both the EC and Japan. The EC 6th Framework Programme includes an approximate $700 million investment distributed over five years in thematic area 3, dedicated to nanotechnology and materials.

Nanobiotechnology is included in this area in two sections: integration of biological and nonbiological systems at the nanoscale level and integration of new nanoscale materials and technology for improved security and quality of life, including tissue engineering, new biomimetic and biohybrid systems. Other EC thematic areas may include some support for nanobiotechnology. Of the approximate $700 million allocated to nanotechnology in the 6th Framework, only a relatively small fraction as compared with the US will be awarded to nanobiotechnology owing to competition with other areas.

Nanotechnology has been formally recognised as one of the science and technology priorities by the Federal German government in FY 2003, with a budget of about $118 million (assuming $1 ¼ 0:95 Euro). This programme was announced in May 2002. The focus is on nanoelectronics, nanoscale materials, and optical science and engineering. A lower importance is given to nanobiosystems, an area which is supported in a Virtual Nanotechnology Competence Center located in Munich. In FY 2002, the UK had a budget of about $45 million allocated to nanotechnology, distributed throughout various government programmes. While the areas of strong specialisation are nanoelectronics, nanophotonics and molecular nanotechnology, the contribution of medical and pharmaceutical research is estimated to be about 5-7%.

France had a FY 2002 investment estimated at $35 million ($50 million planned for FY 2003) for nanotechnology under a variety of programmes, and the contribution of research on nanobiosystems is estimated to be in the same range of 5-7%. Australia has created a national R&D programme on nanotechnology including a focus area on biomimetics and biosensors. The Korean government has designed nanotechnology as one of sixR&Dnational priorities, and in July 2001 announced the nanotechnology priority programme of $1. 2 billion for 10 years.

Then, a decision was taken to accelerate the investment from about $87 million in 2001 to about $170 million, of which about $28 million was allocated to a centralised nanofabrication facility and $7 million was allocated to education. The focus is on areas with highest commercial potential and competitiveness over industrialised countries. Nanobiotechnology has not received significant funding in Korea because of existing expertise and higher economical expectation for electronic and materials sectors. The formation of the Taiwan National Nanotechnology science and technology Priority Programme was announced in September 2002 and began in January 2003. The programme has a total budget of about $110 million/year for 6 years.

One of the five research topics is nanoscale biology, with a focus on 'molecular detection and manipulation, bio and non-bio interface devices and materials, and drug delivery devices'. China has an increased budget for nanotechnology estimated to be about $80 million from the central government (and $200 million including regional governments) in 2002. The focus is on nanoscale materials and devices and there is a relatively small contribution to biosystems. Nanobiosystems is an area of interest recognised in international studies on nanotechnology prepared by both the Asia-Pacific Economic Cooperation (APEC) and the Organisation for Economic Cooperation and Development (OECD).

In a survey made by the UK Institute of Nanotechnology, experts identified the location of the most sophisticated nanotechnology developments in the 'Medical/Pharmaceutical area' in the US (48%), UK (20%), Germany (17%), Switzerland (8%), Sweden (4%), and Japan (3%). In FY 2003, the US NNI plans to devote about 12% of its R&D budget to nanobiosystems, Germany about 8%, and France about 6%. The biological route to nanotechnology may be the pathway of choice for countries with less developed economies, because the research facility investment is lower.

The NNIstrategy is to support five complementary modes of research: fundamental research, grand challenges, research centers of excellence, fabrication and user facilities for developing physical infrastructure, and education and societal implications. The main components of other national programmes have similar structures. The US, Japan, Korea, China and other countries have established coordinating nanotechnology offices at the national level. The NNI was a bottom-up initiative that has advanced relatively quickly from its concept to implementation, because of a bold scientific vision and strong societal relevance. Fundamental research for biosystems has been recognised in all phases of the programme.

The NSF listed 'nanobiosystems at the nanoscale' as the first theme in the programme solicitations for FYs 2001-2003. NRC reports on NNI 'Small wonders - Endless frontiers' and 'Implications of emerging micro and nanotechnologies' recommends research on biologically inspired materials and systems, and particularly research on potential applications for sensors, communications and improving human performance. The second mode, known as the 'grand challenges', supports specific visionary R&D activities identified as having the potential to realise significant economic and societal implications. Centers and networks can encourage the multidisciplinary research that is expected to contribute highly innovative approaches to new technology.

Fifteen new multidisciplinary centers have been initiated throughout the country since 2001. Eight were funded by the NSF, three by the DOD, and four by NASA. Fabrication and user facilities for physical infrastructure includes the National Nanotechnology User Network (NNUN) and the Network for Computational Nanotechnology sponsored by NSF, and a group of five large userfacility Nanoscience Centers established by DOE. Of these, the Lawrence Berkeley National Laboratory and Oak Ridge National Laboratory will have a focus on nanobiosystems research.

In 2003, the NSF has opened the competition for the National Nanotechnology Infrastructure Network for a $14 million/year investment for 2004-2013 to replace and expand the goals of the NNUN. It is planned that biosystems research will be a central research and education area in the new network. The NNI provides funding for research that addresses the ethical, social, economic, and workforce impacts of nanoscience and nanotechnology. Examples of funded research areas include transforming individuals and social institutions, identifying impacts of legislation and regulation, reducing barriers to nanotechnology diffusion, and using nanotechnology for more effective education.

Education and training outreach activities must begin in primary and secondary schools. The National NanotechnologyCoordination Office, the secretarial office ofNSET, has prepared a paper entitled, 'Extending Outreach Success for the National Nanoscale Science and Engineering Centers - a Handbook for Universities', which provides guidance for impacting the curricula development process. Broader nanobiotechnology recognition is sought through outreach to the public through the media and museums, secondary and continuing education. Increased interdisciplinarity in university education has been recommended to ensure a nanotechnology workforce.

Ethical and moral concerns also need to be addressed before new developments can be made in some areas, for example, neuroethics need to be investigated before brain and neural systemresearch. Research and education on nanobiosystems is expected to increase within the nanotechnology programmes as our understanding of complex nanosystems develops and new tools become available.

11

New Biocatalysts

Chemical biotechnology is the rapidly growing application of biotechnology to chemical production. It often goes hand-in-hand with green chemistry and the use of renewable feedstocks. Other applications of biotechnology lead to new products, new manufacturing methods and improved economics with lesser demand for energy, or fewer negative or deleterious impacts on the environment. Chemical biotechnology has made a big impact in the industry structure as firms have been acquired, divested, and restructured around various biochemical innovations.

Biotechnology is very pervasive in the food industry - e.g., enzymes for starch manufacturing, beverage production, meat preservatives, etc. Many more impacts in the food industry are certain to come from use of biotechnology. A high level of uncertainty exists in biotechnology relative to recombinant foods and the impact of genetic engineering on crops, but much of this uncertainty is being addressed with facts and information that demonstrate food safety and safe practices in farming in the U.S. The perceptions abroad are not as favorable.

Pharmaceutical use of biotechnology is virtually assured; even though policy issues related to biomaterials, textiles and similar products have yet to be addressed, these areas seem to be on a reasonably firm base. In the past, initial reactions led to great caution in applying recombinant DNA technology to real world problems. This caution has subsided as understanding of the technology's benefits has increased. Similarly, the use of biotechnology in industry may go through a cautious phase only to blossom as a new, and even more widely accepted, positive force for mankind. Indeed the history of the entire Industrial Revolution may be looked upon as a model for the future of the biotechnological revolution.

Visioning Process and Drivers

This strategic plan or "roadmap" is one of several that have or will be prepared to address key areas of research in the chemical industry. This contrasts with the visioning process in other industries, e.g., aluminum or glass, where a single visioning process and a single roadmap may easily suffice to describe the most relevant research needs, and barriers, in the industry. The difference, of course, is that the chemical industry is very unlike such monolithic industries as aluminum or glass. Indeed, "chemicals" consists of both small companies with a limited number of products and a smaller number of large multi-national companies with hundreds of products, processes and intermediates.

The relationship between chemical companies is often symbiotic with suppliers supplying end users, intermediate producers or even their own suppliers. Some companies even integrate their feedstock supply, production, and use functions. However, the greatest majority of companies operate in the horizontal industrial mode and maximise their capabilities and strengths in their own niches in order to produce a slate of products which they can sell. Visioning for the chemical industry is further complicated by the problem of definition. Should petrochemicals be included? Are agriculturally derived chemicals a product of the chemical industry?

And while pharmaceuticals are really "fine chemicals," the nature of pharmaceutical markets and the financial behemoth they represent often cause analysts to consider them separately from fine chemical manufacturing, such as that used in dye making or as fuel or plastic additives. Definitions. The definitions of "chemical research," "biotechnology," "chemistry-based industries," and even "the chemical industry" can be, and are, hotly debated. No matter how we define these areas now, it is agreed by many that the changes that we will see in the next two decades will so revolutionise our industry that it will not seem recognisable to a visitor from the year 1999.

The maturation of technology from scientific breakthroughs to commercial application follows the same kind of growth as microorganisms. The long research and development cycle leads to implementation at small scales followed by large scale, mature production. Biotechnology, as we are seeing it applied today, had very few applications after the seminal discoveries about 1974. In the 1980s, the health industry, particularly pharmaceuticals, found value in the technology and began to employ it extensively.

Now several biotechnology produced pharmaceutical products are entrenched in production. Large-scale commodity food chemicals such as ethanol, high fructose corn syrup, citric acid, and amino acids also employ microbes or enzymes. However, the inroads into commodity chemical production lags far behind these other chemical applications. Therefore, in an overall sense, biotechnology, in many aspects, is still in its infancy and growing like a teen-age youth in other areas.

It is clear that the use of biotechnology has not been fully realised. Even so, the industry still boasts some very impressive facts vis-á-vis societal and financial impacts. In 1999, the biotechnology industry accounted for $13.4 billion in sales and $18.6 billion in revenues. There are over 1,280 biotech companies in the United States. The industry represents a $97 billion investment market, employs more than 153,000 people in high-wage, high-value jobs, and accounts for over $9 billion in research and development. There are more than 90 biotechnology drug products and vaccines helping more than 200 million people worldwide. Biotechnology is responsible for hundreds of medical diagnostic tests.

Biotechnology has brought nutrition and health improvements to foods such as corn, soybeans, tomatoes, carrots, and peppers. Biotechnology has augmented the efficient and effective clean-up of hazardous wastes and spills. DNA fingerprinting is a biotech process which has dramatically improved criminal investigation. In the health-related fields where biotechnology has had its greatest commercial impact, cause-oriented groups tend to raise fewer problems - everyone is in favor of technology that delivers life-saving therapies. Thus the development of cancer fighting "biologics," i.e. protein and nucleic acid therapeutics, has been widely heralded.

Of importance to the "chemical enterprise" and chemical research is that this acceptance has created a commercial incubator for techniques that are being and will continue to be applied to, and transform, even the "core" chemical industry. These successes, combined with advances in the basic sciences underlying biotechnology, are already being felt. These include advances in aspects of the industry that are commodity-driven, ranging from soaps and detergents to textiles and "synthetic" fibers. These advances also embrace "green chemistry," which is driven by energy and environmental concerns.

As energy and/or environmental factors, including global warming and the carbon cycle, are increasingly emphasised, it is hard to see how any part of the industry will be unaffected, from desulfurisation of fuels to bio-remediation of wastes, from steel production to ore refineries. Such optimistic and glowing generalisations must, of course, be backed by concrete examples. The literature, e.g. Trends in Biotechnology, Biotechnology Progress, or Biotechnology & Bioengineering (the highest impact factor journal in biochemical engineering for the past three decades) and international journals such as Nature Biotechnology, is replete with promising ideas to replace traditional synthetic chemistry with biotechnologybased chemistry, e.g., "plants as factories," combinatorial phage display, and abzymes, synzymes, and "newzymes" (enzymes altered to have unique catalytic properties).

Already the term "metabolic engineering" is replacing "genetic engineering" and "functional genomics" is replacing "genomics" as common buzzwords. Chirality is all-important in the field of health-care products, and enzymatic chiral resolution, although by no means new, is a prime application for new biocatalysts.

Driving Forces for Biocatalyst Technology

Four major forces drive biocatalyst technology in "chemical-based" industries.

Societal forces: Society is constantly demanding new technologies, new products, and new ways of living. Examples include health products such as biosensors, tissue plasminogen activator (TPA), and Epogen®, and commodity items such as stain-removing detergents, preworn-looking blue jeans, and reduced-calorie food products. Society also demands that technology development includes minimal environmental impact.

Business forces:- Profit/cost reduction drives many of the new changes. Will a biocatalyst give the same, or better, product at a lower cost? Are there good solutions to the problem of the high cost of isolating bioproducts in downstream processing? Active efforts in bioseparation systems are being undertaken to drive down the cost of biocatalysts to make them competitive with classic inorganic/organic catalysts. Also, business and economic demands are now requiring that better technologies be developed for selection and production of relatively cheap enzymes.

Government-, regulatory-, and "cause"-driven forces: Concerns about "greenhouse gases," especially CO_2, will drive new "closed carbon cycle" methods, e.g., fine organic chemicals produced by metabolic engineering of crops. High laurate canola and corn are examples of how agricultural approaches to using whole organisms as biocatalysts create a product that would be energy intensive and likely yield more greenhouse gases if it were produced from petroleum (this postulate is still being evaluated by life cycle studies but generally appears to be true). The issue of genetically modified organisms will likely lead to regulatory pressures to completely characterise biocatalysts for efficacy and safety. Government policies already favor the use of biocatalytic processes for producing fuel ethanol, but such policies could change.

Basic research pressures: Significant industrial and practical technologies result from the search of basic science for truth and discovery. These "just because it's there" discoveries often lead to huge advances or potentially impactful advances. Examples include polymerase chain reaction (PCR) technology, which was designed for multiplying minute quantities of genetic information for research purposes but is now employed to produce useful genetic libraries for plant, animal, and microbial species or for forensic diagnostics.

DNA sequencing was designed to help researchers decode genetic information, but with the advent of improved methods, the sequencing of the human genome and other important crops and species could lead to new health therapies or improved crop production. Genetic engineering was initially formulated out of a desire to understand how to transfer genetic information amongst similar microbial species. Now it is used to produce life-saving therapeutics such as insulin or human growth hormone, but can also be used to increase milk production in cows, manufacture new polymers, or develop new therapeutics. The following recasts these drivers in an outline that displays some of the specific aspects of each of the drivers.

Biocatalysis has a large impact in the chemical world. The enzyme market alone is a $1 billion global business. Traditionally, microbial and enzymatic processing have been used to convert biologically-derived (or renewable) feedstocks. However, they are increasingly being used with materials derived from fossil fuels. Uses are as divergent as chiral enzymatic transformations within an organic synthesis for a drug or for microbial desulfurization of diesel fuels. It suggests the potential and challenge of applying biocatalysis to all feedstreams.

Since pharmaceutical manufacturers are among the major customers of fine chemical manufacturers, the synthetic organic chemical industry will be impacted and transformed by the revolution in biocatalysts. Examples include production of precursors for Glaxo Wellcome's HIV drug, Ziagen, and DSM's Fine Chemicals production, including intermediates for aspartame, amoxicillin, and a variety of classical pharmaceuticals, e.g., diltiasem and captopril.

The chemical process industries are beginning to realise that enzymes are not only effective for catalysing reactions of "natural"compounds within living systems, but that they can be used to catalyse reactions of "unnatural" compounds. Enzyme biocatalysts are being applied in the production of fine chemicals, pharmaceuticals, and agricultural chemicals. Their attractiveness comes from high selectivities, ability for use under ambient conditions, and ease of disposal. The enzyme nitrile hydratase from a R. rhodococcus strain has been developed for the hydrolysis of acrylonitrile to acrylamide for use in plastics. The enzyme is immobilised in whole cells and can produce acrylamide concentrations greater than 600 g/L.

The biocatalytic approach has reached a production level of 100,000 tons/yr. The DSM-Toyo Soda process uses the enzymatic protease thermolysin for manufacture of aspartame, and is illustrative of two types of biocatalyst selectivity: chemical and stereoselectivity. High-fructose corn syrup produced in large quantities (23 million tons in 1998) is an enzyme-based product. The process includes three enzymatic steps: the -amylase catalysed liquefaction of corn syrup, further hydrolysis of sugar oligomers by glucoamylase, and the isomerisation of glucose to the glucose-fructose mixture.

The hydrolysis of penicillin G or V to 6-aminopenicillanic acid (6f-APA) using penicillin acylase is an early success story for the use of enzymes in chemical manufacture. Almost 9000 tons of 6f-APA were produced worldwide in 1995, mostly via the biocatalytic approach. DuPont and Genencor have filed patents for processes and microorganisms to make 1,3 propanediol (1,3-PD) by fermentation in one step from various carbohydrate sources. The 1,3-PD is used in the production of the polyester polytrimethylene terephthalate.

Cargill-Dow Polymers is developing a large-scale fermentation process alongside their other corn processing systems followed by chemical processing (a type of biorefinery) to generate polylactic acid for a multitude of applications including biodegradable sutures, biocompatible fibers, packaging, and functional replacements for commodity plastics such as styrene. Many researchers in academia, industry, and government are developing directed molecular evolution to produce enzymes with novel functions.

The technique uses mutagenesis and DNA shuffling to generate random mutations in the genes of interest. This approach can increase the activity of selected enzymes by more than 100-fold over the native enzyme. Novel enzymes derived from extremeophiles represent an area of high research and commercial interest. These enzymes have gained attention because they evolve under circumstances that can provide activity over a broader range of conditions. Enzymes (native or evolved), along with microbial catalysts, are used combinatorially to discover new biologically active molecules or to improve lead candidates for pharmaceutical discovery.

Biocatalysis can have a broad impact in multiple aspects of the chemical industry. Biocatalysts can offer multiple advantages as described above, but to gain these advantages, biocatalysis and its application bioprocessing need to become a predictable and "routine" tool for conversion. The development of biocatalysis should be seen as adding processing and catalytic tools to the more traditional chemical and thermochemical methods of converting one chemical or material into a new and more valuable material. We need to find ways to expand the utility and impact of biocatalysts by eliminating the typical difficulties or operational limits encountered in biocatalysis, such as temperature, pH, product inhibition, slower rates, and processing dilute aqueous product streams. This must all be accomplished while maintaining the advantages of high specificity and multistep processing.

High specificity allows precise reactions to occur (e.g., removal of organosulfur from diesel). Similarly, multistep processing improves yields both by elimination of intermediate synthetic steps and by decreasing solvent switching, thus reducing the number and volume of waste streams. The bioconversion industries are in an explosive stage of development. Much of the root cause for this current state of excitement has been technology-driven. The technology has pulled the markets into new areas. In particular, with the more widespread use of biotechnology tools, the health industries have now become largely market-driven.

Moving the larger scale bioprocesses from a technology pull to a market-driven basis is a continual challenge. Achieving market pull in the commodity chemical business will require new approaches and tools, a track record of reasonable successes and, in many cases, the creation of new markets. The formation of CargillDow Polymers, LLC is an example of a joint effort between the producer of corn syrup sugar-based lactic acid, Cargill, and the product development company, Dow Chemical, which is devising methods to convert lactic acid into commercially viable products.

DuPont's plans for 3-carbon polymers based on 1,3 propanediol includes the temporary use of a pure chemical process to establish the market while a biocatalyst-based process is commercialised. These ventures indicate that the biocatalysis area is moving to a market-driven mode that actively pursues and seeks out new technologies for exploitation rather than waiting for them to be offered.

How do we reach these goals? One perspective views biocatalysts as just another catalyst possessing commercial perspectives of being "faster, cheaper, better". This perspective requires that the R&D of biocatalysts be directed towards reducing the time to market and implementation. This perspective demands implementation of recent biotechnology tools as well as development of new innovations.

The goal of the roadmapping effort was to acquire sufficient information and input to craft an R&D pathway for the development of biocatalyst technology for applications in the broadest definition of the chemical industry. This definition includes the production of commodity, intermediate, and fine chemicals, ranging from basic chemicals to pharmaceuticals and using both conventional and unconventional feedstocks. Conventional implies petroleum or coal-based feedstocks.

Unconventional includes renewable biomass, natural gas, using gases such as carbon dioxide or nitrogen, other inorganics, and feedstocks from marine environments (i.e., sulfur compounds). Sufficient information was obtained to initiate the construction of a first-generation, biocatalysis roadmap. As R&D progresses, specific performance indicators will appear that will serve as markers of true progress.

Use of biocatalysts by process chemists will become as routine as the use of conventional catalysts. Biocatalysts will be viewed as reagents that can be obtained from a catalog (some are today) or ordered up from a biocatalyst development company. The range of environments for biocatalyst application will expand beyond simple aqueous systems into nonconventional, complex multiphasic or multimilieu environments. Biocatalyst function will become more predictable, particularly with respect to reactions catalysed under the aforementioned conditions.

Biocatalyst development time (from design to synthesis to deployment) will become competitive with conventional catalysts. Functional platforms based on substrate, milieu, reaction, etc. will become available, from which it will be possible to rapidly screen and develop biocatalysts for tailored use.

Structure/function properties of biocatalysts will become more available. This will facilitate wider applications in industry for a wide range of biocatalyst substrates or feedstocks.

The strategy has two components:

1) promoting R&D in biocatalyst research and use, and
2) monitoring the progress of biocatalyst development and deployment.

Promoting R&D Support: Steps will be taken to encourage federal, private, and nonprofit funding for biocatalyst development in support of the R&D identified in the roadmap. Areas of R&D interest and opportunity include:

a. Conventional processes (aqueous processing on small to medium-sized scales) for typical industries such as specialty and fine chemicals (pharmaceuticals) and in unconventional arenas such as commodity chemicals and petrochemical processing (organic media, gas phase, etc.). Organic media should not be specific to commodity and petrochemical processing, rather, it should be examined in all potential arenas.

b. Biocatalyst discovery and the need for additional resources for the design of new biocatalysts with optimised/tailored properties.

c. New avenues of biocatalyst use for processing renewable biomass resources, renewable gas conversion (carbon dioxide, nitrogen oxides, etc.), and other unconventional resources, e.g., inorganics.

Included in this component will be the development of biocatalysts in unconventional hosts such as plants, animals, etc.

d. Tool development and fundamental understanding of biocatalyst structure and function. This will provide the critical technology to allow biocatalyst developers to make significant advances in the other two strategies.

Specific approaches to include more biocatalyst research in prorammes and R&D calls will require some tailoring of the information in the documents to the target audience among which will likely be industry, academia, and federal agencies. The approach for R&D organisations employing a prorammatic plan to solve mission-oriented problems will require targeting the organisational proramme managers, the actual performers of the research of that programmatic effort, and their stakeholders.

For R&D organisations employing a solicitation process, the target will be those who craft such solicitations and the approach will be to have these solicitations reflect the portions of the roadmap that can logically and rationally be integrated into their solicitation(s). For R&D organisations with the mission of supporting principal investigator-driven, peerreviewed work, the targets will likely be the academic and basic research community itself. This will require exposure of the roadmap concepts in public forums. Several industries that use biocatalysts, including chemical, agricultural, and forest products, have encouraged some federal funding agencies to support work in precompetitive research in areas with long-term benefit to their industry.

R&D Monitoring: A successful and productive implementation of the roadmap by research institutions will require that R&D recommendations be monitored for use within the research community and, in particular, within the chemical industry. This can be accomplished by tracking federal and private funding for biocatalyst R&D, new industry developments in bioprocessing, and new biocatalyst citations in the literature.

A two-fold approach is recommended to accomplish the above strategies. First, ownership for promoting the R&D identified within the roadmap will be determined. Second, mechanisms will be established to track and monitor biocatalyst R&D and deployment. Two areas that this approach will address include a trained work force and the current specifications of industrial processes. Many chemists and chemical engineers are still unfamiliar with biocatalysts and enzymes. This is changing but will not occur overnight.

The approach involves educating this group of influential stakeholders. Also, 90% of commodity chemical processes involve organic solvents which is not the typical milieu for biocatalysts. In addition, these processes normally operate at temperatures and pressures much higher than is suitable for biocatalysts. This approach will seek to make inroads addressing this 90% market and also to enlarge the 10% of the market where water is present and biocatalysts are more likely to be active participants.

The approach involves all R&D sectors in the United States - industry, not-for-profit, academia, and government. The target audiences include all agencies that support biocatalyst R&D, industries and companies involved in biocatalyst development, and non-governmental organisations that would have an interest in the development of biocatalysts, especially organisations concerned about genetically-modified-organisms (GMOs).

Ownership. There is a need to identify a lead organisation and to establish a mechanism for the interaction between the lead and other stakeholders in this R&D arena. One approach is to form a biocatalyst executive steering group commissioned to further the strategies and objectives of the roadmap. Representatives to the steering group could be from other trade organisations or appropriate industry, academia, and laboratory personnel.

The Council for Chemical Research (CCR) has been active in support of the various Vision 2020 roadmaps and will be encouraged to establish a central role. The other signatories of the Vision 2020 compact will be enjoined to participate with roles commensurate with their interests and the interests of their stakeholders. For example, the American Chemical Society, Council for Chemical Research and the American Institute of Chemical Engineers might focus on promoting the roadmaps to academic, national laboratory, industry and other research institutions they each represent. The Chemical Manufacturer's Association and the Society of Organic Chemicals Manufacturing Association might represent corporate manufacturing interests. Additional representation will be sought from the Biotechnology Institute Organisation (BIO), which represents the biotechnology industry and other trade and not-for-profit organisations.

A key to successful implementation of the roadmap is tracking R&D progress and monitoring R&D results. Biotechnology/ biocatalysis funding through various federal agencies and departments, papers published in research journals, corporate R&D investments, venture capital investments, new company developments, new bioprocess applications, etc., are areas to monitor with respect to biocatalysis R&D progress and the general business health of biotechnology related to biocatalysts. Various for-profit, not-for-profit and federal organisations with experience in this type of tracking and monitoring work include the Rand Corporation, Biotechnology Industry Organisation (BIO), the National Science Foundation (NSF), and others.

Benefits of Roadmap Information

The use of the roadmap by funding agencies, industry, and other R&D organisations should provide some increased awareness of both the challenges and opportunities that lie within the realm of biocatalysis. The introduction to this roadmap depicts the state-of-the-art and the ongoing R&D efforts in this field. The consensus that the field is ripe and that opportunities abound to harvest a rich supply of useful biocatalytic tools is widespread. The user of this roadmap will benefit from the broad experience of a multitude of scientists in the various fields of biocatalysis. This roadmap can be a powerful tool in crafting requests for proposals, in outlining research prorammes, and in addressing specific biocatalytic development pathways.

12

Evolution and Regulation of Pharmaceuticals

Biotechnology industry was formally established in latter part of the 20th century. This technology has numerous applications. Mainly it is used for the production of food, medicine and other goods for thousands of years. Biotechnology has its origins in the collection of wild seeds for cultivation and the domestication of some species of wild animals by early agrarian societies. Farmers selected seeds from the best plants to be sown again the next year and bred only their finest animals.

Through selective breeding, humans developed new plant and animal varieties that were most beneficial to early societies. Humans also discovered how to use enzymes and microorganisms to produce commodities like wine, beer, cheese, and bread. Since little was known about the science behind these processes, only skilled craftspeople practiced the arts of baking, brewing, and cheese making.

The first clearly recognisable sic application of modern biotechnology in the Western world was probably the use of cowpox to immunise sic humans against small pox by Edward Jenner in 1797. Doctors continue to use attenuated organisms to build up immunity against a more dangerous organism to prevent the spread of many illnesses such as measles, chicken pox, and the flu. This method was coined vaccination since the less harmful microbe in Jenner's experiment came from cows or vacca in Latin. A new discipline, molecular biology, emerged in the late 19th and early 20centuries when scientific research began to focus on heredity, genes, and proteins.

Alfred Hershey and Martha Chase established that deoxyribonucleic acid (DNA) is the genetic material in 1952. A year later, James Watson and Francis Crick developed a structural model of DNA. After this discovery, research on cell function and replication quickly followed. In the 1970s, "the experiments of Paul Berg, Herbert Boyer, Stanley Cohen, Janet Mertz, and Ronald Davis and their colleagues... ushered in the era of modern biotechnology and genetic engineering". This group revolutionised the field of recombinant DNA technology.

DNA from two different sources could be combined and transferred to a host cell, which then would display a new genetic trait, such as antibiotic resistance or the production of insulin. Around this same time, monoclonal-antibody technology was also discovered, which is an important medical diagnostic tool. With the advent of these new technologies, safety and regulation concerns arose.

By 2001, the worldwide industry was represented by over 4,200 public and private biotech companies, concentrated in four key markets-the United States, Canada, Europe, and Asia/Pacific. The U. S. led the biotechnology industry with over 70% of global biotech revenues and research and development expenditures. Canada had the second largest number of biotechnology companies, numbering over 400. Europe's biotechnology market is continuing to grow with more product approvals and venture capital financing than ever.

The Asia/Pacific market is emerging and expanding across the region with Australia, China, India, and Singapore leading the way. Despite the economic downturn in 2001, the global biotechnology sector experienced its second best financing year ever and its stocks outperformed high-tech and blue chip sectors. According to Ernst & Young, one factor contributing to this growth is consolidation in the biotechnology industry that has been taking place in order to create more product-oriented companies with potential for long-term growth.

A second factor is the late-stage venture investment being made so that emerging companies can more fully develop product candidates and their prospects for success before going public. As the biotechnology industry has become more geographically diverse, companies have pursued more cross-border collaborations. Many governments have recognised the potential of the growing biotechnology sector and are encouraging expansion in their countries by devoting government funds to the industry. Global competition in the biotechnology industry has also increased as investors search for the best deals and most promising technologies, regardless of location.

Biotechnology and Pharmaceuticals

Recent advances such as those in genomics have quickly found their way into the pharmaceutical industry where they are helping the development of new drugs and therapies. More than 250 million people have benefited from medicines and vaccines developed through biotechnology. Nearly 200 diseases are being targeted by research conducted by 144 companies and the National Cancer Institute. Success in the biotechnology drug industry, however, is infrequent due to the small rate of product approval.

For example, while 24 biotech drugs, vaccines, and new indications and uses were approved in the U. S. in 2001, there were over 350 still in the clinical trial phase. A company's ability to successfully bring a biotech drug to market can be critical for survival. When the U. S. Food and Drug Administration failed, for example, to approve the ImClone drug, Erbitux, in late 2001 the company's shares fell dramatically. ImClone had a late stage clinical alliance with the pharmaceutical company Bristol-Myers Squibb to provide for the development and commercialisation of this humanised monoclonal antibody to the epidermal growth factor receptor for the treatment of cancer. A recent trend in the biotechnology drug industry, clinical trials are often undertaken in collaboration with large, established, and experienced pharmaceutical companies..

In the past, biotechnology and pharmaceutical companies primarily used licensing agreements to facilitate collaboration. Licensing agreements are used when the main purpose is to grant pharmaceutical companies intellectual property rights for products and processes developed by biotech firms. The number of biotech-to-biotech mergers and strategic alliances has, however, been increasing as the biotechnology industry continues to expand its focus, and size and expertise in many fields are necessary for success. Biotech companies and pharmaceutical companies are now entering into early stage collaborative ventures so that both firms share development costs, risk taking, and profit sharing.

The biotech companies can also improve their reputations and validate their technologies by aligning with pharmaceutical companies. For example, Millennium Pharmaceuticals made more than 20 alliances with big pharmaceutical companies between 1995 and 2001, giving it more than $2 billion in funding, while it raised another

$830 million in stock sales. More recently, Millennium has shifted from using licensing agreements to forming collaborative ventures with companies such as Aventis and Abbott Laboratories. Both companies share costs, risks and potential profits and Millennium maintains significant ownership and decision-making rights from discovery through commercialisation.

In April 2002, Millennium purchased Cor Therapeutics, making it a company that has become a fully integrated biopharmaceutical producer with two acquired products on the market and another 10 in clinical trials. While Millennium expects to earn $300 million in revenues this year, it is not yet a profitable company. Research and development costs have increased and two of its drug candidates have had disappointing clinical trials. In addition to collaborations between biotechnology and pharmaceutical firms, governments support a variety of programmes to promote research and development in academia and the private sector.

In the U. S., the Advanced Technology Programme (ATP) of the National Institute of Science and Technology (NIST) has awarded more than $40 million to small biotech companies. The Small Business Innovation Research (SBIR) Programme awarded more than $300 million in grants in 1999 to support collaborative agreements between the NIH and private biotechnology companies. Asian governments are also supporting their biotechnology industries. India's Department of Biotechnology has set up a $21 million venture fund. Japan has allocated $2. 4 billion for science and technology over the first half of this decade.

Singapore has set a goal to become the center of biopharmaceutical research and development in the region. Taiwan has been promoting biotechnology as the next major economic growth sector. In Europe, the United Kingdom has the largest number of biotechnology companies and the annual government expenditure on bioscience research is approximately £650 million. Germany has been working to catch up with the UK in the biotechnology sector through the Biotechnology 2000 Programme with an annual budget of £124. 5 million. Germany's Bio Regio contest has sought to support cluster growth in Munich, Rhine/Neckar and Rhineland, and stimulate other regions to create initiatives of their own.

Seventeen regions participated in the BioRegio competition and three received awards of DM50 million over five years. France and the Netherlands are also pursuing programmes to expand their biotechnology industries. The Dutch government provides life sciences academia with $50 million to support the formation of new companies. Private investors are also supporting biotech companies, now on a more global scale. Venture capitalists are investing in companies outside of their local markets. Between 1995 and 1999, the share of the biotechnology sector in total venture capital investment grew at an annual average of 24% in the U. S. and 37 % in the European Union.

The EU and Canada provided more support to the biotech sector in relative terms. Canada, Belgium, and Switzerland respectively allocated 0. 23, 0. 21 and 0. 14% of GDP to the sector as opposed to 0. 13% of GDP in the U. S.. Although expanding, the biotechnology industry is still relatively concentrated in a few countries. During 1999, more than 80% of U. S. biotech exports went to seven countries in the Organisation for Economic Change and Development. The remainder of the biotech exports went to other countries within OECD. In 1999, Belgium was the leading destination for U. S. biotech products at 24% of exports. Japan, Germany and Canada were also significant recipients of U. S. biotech exports. Belgium was also the leading supplier of biotech products to the U. S. at 26%, making it the main partner for the U. S. biotech industry during the late 1990s.

France and the Netherlands exported more biotech products than they imported from the U. S.. Switzerland and Ireland also exported a significant number of biotech products to the U. S, but imported very little. Although the majority of biotechnology has been taking place in a few countries, it has the potential to spread throughout the world in the next few years in places such as India, Australia, and Taiwan. Not surprisingly, private sector activities in the biotechnology industry are beginning to organise around technologies that have near-term payoffs rather than technologies that will take more time to develop.

For example, bioinformatics has proven to be profitable and is receiving more attention. On the other hand, stem cell technology and gene therapy have yet to result in commercial applications and have been receiving less funding from investors. Investors are willing to take more risks and fund a variety of projects when the market is good, and are less willing to take risks when the market has taken a downturn. Research and development in the biotechnology sector is proving to be more demand driven than it had been in its infancy. Because of changes in the biotechnology industry and the slower-than-expected pace of new drug development, pharmaceutical companies are beginning to realise that they will need to make some radical changes in their operations if they are going to remain competitive and innovative.

Goals envisioned in the late 1990s have yet to be realised. Pharmaceutical executives had anticipated that by the year 2000, the time to bring a drug to market would have been halved; the number of new molecular entities entering clinical development would have been tripled; attrition of drugs in the late stages of development would have been reduced; and growth objectives by focusing on therapies with "blockbuster" potential would have been achieved. Instead, companies have fallen short of meeting these goals, the cost of research and development has continued to rise, and commercialisation expenses are increasing.

Pharmaceutical companies are also being pressured by markets to meet high growth projections within the next decade. The original goals were based upon the expectation that researchers could identify valid drug targets by examining gene sequence information. Because this has not occurred, however, a significant time lag between genomics and drug discovery remains. Gene sequence does not reveal much about protein function or relevance to disease, so the true value of a genome sequence will be realised only after the function of the protein or proteins that it encodes for have been discovered.

Therefore, the biotech industry has recently been turning its attention to proteomics, which is the analysis of differences in protein characteristics induced by different disease states. This could lead to the discovery of ten times as many targets as genomics. Unfulfilled expectations followed by yet wilder achievement predictions are known to scholars of technology/industry evolution as characteristics of areas with high technological opportunity. This group certainly includes contemporary biotechnology that is revolutionising the way we feed and cure ourselves.

Changing Ways of Pharmaceutical Production

In the past, much of drug discovery relied on trial and error and in some cases luck. One of the most famous pharmaceutical discoveries of the 20century occurred through almost sheer serendipity. During the 1920's

Alexander Fleming had been looking for better antiseptics. "While cleaning his laboratory one afternoon, he noticed large yellow colonies of mold overgrowing a culture of staphylococcus bacteria on an agar plate. Fleming realised that something was killing the bacteria, and he proceeded to experiment with juice extracted from the mold by spreading it on agar plates covered with more bacteria. He found that even when the juice was highly diluted, it destroyed the bacteria". He called this new substance Penicillin. In the late 1930s and 1940s researchers began to use a more systematic approach to look for medicinal products. They also searched for naturally occurring compounds in plants and other organisms.

When a natural product was found to be effective for treating disease, chemists would then attempt to synthesise the compound out of more abundant resources. For example, quinine was isolated from cinchona, an evergreen tree native to the mountains of South and Central America. Because collecting natural quinine was an expensive and strenuous task, William E. Doering and Robert B. Woodward synthesised quinine-a complex molecular structure-from coal tar. Researchers also began to synthesise substances that naturally occur in the human body to help treat diseases. Through the years, researchers began to have a better understanding of how DNA and proteins functioned. They began to use a variety of forms of chromatography, mass spectrometry, proton nuclear magnetic resonance, and other analytical techniques to discover and analyse new drugs.

Scientists today are using a variety of new techniques to discover and design new medicines in addition to the old approaches that relied on persistence and serendipity. Some of these new techniques include: gene chips; combinatorial chemistry; high-speed computers that identify likely drug targets; laser microscopes that "capture" individual cells; recombinant DNA; and polymerase chain reaction that amplifies a sample of DNA very quickly. With these new technologies, companies can now examine in a single day compounds that in the past would have taken years to evaluate.

The completion of the Human Genome Project will help scientists to identify specific genes, the proteins those genes produce, and the functions of those proteins. "The challenge for pharmaceutical research is to unravel the pathophysiology of human diseases and thus, make it possible to identify targets accessible to drug intervention". Knowledge of the human genome and the proteins it produces will help scientists determine how some diseases are triggered and give them more targets for medicines. About 500 targets are currently known, but the mapping of the genome could result in at least 5,000 genes that are important targets or produce therapeutic proteins, suggesting that only 10% of potential therapeutic strategies have been identified and exploited to date.

A target is considered pharmacologically accessible when an organic molecule can regulate the target's function. The target for most current drugs is the G protein coupled-receptor family of molecules. Additional targets include other cell surface receptors and molecules, proteases, protein kinases, and phosphatases. Most of the drugs in development for these targets are therapeutic proteins and monoclonal antibodies and are directed toward cancer and inflammation targets. Scientists are using bioinformatics to harness the knowledge gained through the study of the genome and the proteins it produces. Gene and protein sequences are stored and analysed using databases and computer programmes.

Bioinformatics tools vary from standard gene sequence databases such as GenBank which is maintained by NIH, to complex fuzzy logic systems such as GRAIL, which was created at Oak Ridge National Laboratory.

These new computational capabilities can help decrease the amount of time it takes to generate a new drug compound. Libraries of genes, proteins, and potential drug compounds can be screened virtually or "in silico" to see which compounds fit best with the known or projected 3-dimentional structure of the target protein's active site. Technologies to predict pathway analysis, in vivo response, and Absorption, Distribution, Metabolism, Elimination and Toxicology (ADMET) response are being developed and integrated to further identify compounds that will be most effective for drug targets.

As a result, companies hope that the number of drug targets and lead compounds coming into the pipeline will increase, and toxic or less promising compounds can be screened out at earlier stages in the discovery process. The overlap of biotechnology and information technology is one area where collaboration has occurred. In August 2001, for example, Novartis AG and Compugen announced a partnership to create a complete database that will detail gene expression at the level of RNA, known as the transcriptosome. Compugen will also provide Novartis with a DNA chip design that will represent the transcriptosome database. Researchers will use these gene chips to identify genes that are expressed in various diseases, and Novartis will then develop new drug therapies for these targets.

Vaccines

One area in which biotechnology is having a significant impact is the discovery and production of vaccines. In general, biotechnology allows for the safer production of traditional vaccines. In the past, vaccines were grown in a variety of media including human tissue, chicken eggs and cultured cells from animals. By using recombinant DNA techniques and cell cultures, much larger quantities of purer vaccines can hopefully be produced. Researchers are sequencing complete genomes of bacteria, viruses, and parasites to understand the diseases that these pathogens cause.

The genomes of more than 30 important bacteria and parasites have already been sequenced, and over the next two to four years, scientists expect to sequence the genomes of more than 100 other pathogens. This will result in information on more than 300,000 genes, which may offer sources for developing completely new classes of vaccines or therapeutic agents. Researchers are targeting a wide variety of diseases for vaccine discovery including tuberculosis, meningitis B, malaria, HIV/AIDS, and gastroduodenal diseases.

Tuberculosis has reemerged and has been spreading in both developed and developing countries. Two new types of vaccines have been created as a result of genetic research. Already in the clinical trial phase, one vaccine uses a combination of bacillus Calmette-Guérin (BCG) and a booster consisting of a vaccinia virus that is genetically engineered to produce antigen 85, a protein that is synthesised by Mycobacterium tuberculosis. Scientists are also exploring the possibility of an intramuscular injection of a fragment of DNA encoding a heat-shock protein from a bacterium that is related to M. tuberculosis. It has proven to not only protect mice from the disease but cure those that have been infected as well.

A meningitis B vaccine was developed in Cuba, which is currently being tested in Great Britain and has been licensed to at least a dozen other countries. Cuba is just one of a number of developing countries entering the biotechnology industry. Researchers in India are working on a vaccine for Plasmodium vivax, the primary type of malaria in India. The research was a collaboration between the International Center for Genetic Engineering and Biotechnology in New Delhi and the Malaria Vaccine

Initiative with partial funding from the Bill and Melinda Gates Foundation. The vaccine will be developed by Bharat Biotech International in Hyderabad, India.

Other researchers are working on the Plasmodium falciparum parasite, which is primarily found in Africa. The genome contains 5,000-7,000 genes that encode for proteins that are likely targets of the protective immune response that occurs when a person is infected with the parasite. Researchers have recently sequenced the P. falciparum genome to determine which proteins induce the immune response and hope to create a malaria subunit vaccine, which is much more practical than a whole parasite. The complete genome sequence of the bacterial pathogen Helicobacter pylori strain 26695 has recently been published. Scientists hope to screen the genome for potential vaccine candidates for this pathogen that causes a variety of diseases including peptic ulcer, chronic gastritis, and gastric adenocarcinoma.

The vaccine-oriented company, Aventis Pasteur has used bioinformatic and computational methodologies to select bacterial vaccine candidates. Once a candidate was identified, it was amplified using polymerase chain reaction and then purified. The vaccine candidates are being evaluated using murine models. A promising vaccine for HIV/AIDS has gone into clinical trials in Nairobi, Kenya and Oxford, England. It was created as a result of an observation that some prostitutes in Nairobi develop strong cellular immune responses but do not develop HIV infection despite repeated exposure to the pathogen.

Another new vaccine technology that has been made possible by recombinant DNA techniques is the incorporation of vaccines into potatoes and other fruits and vegetables. The vaccines could be eaten as part of a meal rather than administered through an injection. Plant based vaccines would not require refrigeration, making them potentially cheaper and easier to store; they could be grown or freeze-dried and shipped anywhere. Some disease targets for this technology include cholera, measles, and human papilloma virus-associated with cervical cancer. A recombinant hepatitis B vaccine that has been incorporated into potatoes is currently being tested in a human trial.

The process of incorporating a gene for a vaccine or a therapeutic protein into plants or animals is known as pharming. Researchers are exploring the possibility of using recombinant DNA techniques in cows and other farm animals so that they produce a desired protein in their milk. The protein could then be separated and purified to be used as a drug. The Biotechnology Industry Organisation (BIO) has recently addressed the concern that pharmaceutical-producing plants might cross with plants that are intended for the food supply, and has supported strong regulatory oversight. All BIO members will be allowed to grow the genetically modified crops, but only in areas of the country that are not centers of that crop's production. For example, pharmaceutical producing corn crops will not be grown in the mid-west corn belt.

Biologics

Recombinant DNA technology is being used in the production of additional biologics, such as insulin and human growth hormone. In the past, diabetic patients used insulin isolated from the pancreases of cows and pigs to treat their disease. Animal insulin is, however, not identical to human insulin and diabetics can have adverse reactions to it. Therefore, when the technology to insert a gene for a human protein into a microorganism's genome was discovered, scientists immediately identified insulin production by Escherichia coli bacteria as a candidate for this technology. Licensed for human use in 1982, insulin was the first pharmaceutical to be produced using recombinant DNA techniques. A new treatment for diabetes involving tissue engineering is currently being pursued.

Scientists are experimenting with encapsulating Islet of Langerhans cells, the pancreatic cells that produce insulin, in semiporous microspheres so that one day they can be used in transplants for diabetic patients. The microspheres would protect the islets from being attacked by the immune system while allowing glucose to reach the cells and insulin to be secreted into the bloodstream. This treatment would eliminate the need for insulin injections in diabetics. Recombinant DNA techniques can also be used in the treatment of hemophilia, which is caused by the insufficiency of a clotting factor in the blood, Factor VIII.

Previously, the only sources of Factor VIII were human plasma and blood donations, but fears of HIV and hepatitis in the blood supply encouraged scientists to look for other sources. Today, animal cells are used to produce recombinant Factor VIII, which is purified and free of any viruses. Recombinant erythropoietin (EPO), used to treat patients with anemia, is produced in a similar manner. Patients with anemia have a shortage of red blood cells causing weakness and tiredness, and EPO stimulates the production of red blood cells. Recombinant EPO has been produced since 1989, and is primarily used to treat dialysis patients.

Pharmacogenomics

Another promising area of research that will benefit from the completion of the human genome project is the field of pharmacogenomics. Pharmacogenomics is the general study of all of the many different genes that determine whether a patient will have a good response to a drug, a bad response to a drug, or no response at all. Pharmacogenetics is very closely related field that studies the inherited variations in drug metabolism and response. The two fields are so closely related that the terms are often now used interchangeably. Scientists hope to tailor drugs to patients based on slight genetic differences that can affect the way an individual metabolises medicines.

Companies will begin to create families of drugs for a range of patient subpopulations rather than a single "one-size-fits-all" blockbuster drug. Genotype, a person's genetic make-up, rather than phenotype, the physical characteristics that a person displays, will define these subpopulations. Creating personalised drugs is likely to result in greater safety and efficacy in pharmaceuticals. Success rates in clinical trials will also likely increase since the risk of toxicity and adverse side effects will be reduced. In order for pharmacogenomics to be effective, a diagnostic test must be paired with each drug to determine which drug is appropriate for which patient.

Before a doctor can prescribe a drug for a patient in a specific subpopulation, a pharmacogenetic diagnostic test must be performed to determine which drug will have the best response in the patient. One problem that might arise from the practice of pharmacogenomics is the use of drugs designed for specific subpopulations by patients that are not part of the subpopulation. Using a drug in a purpose or manner that is not described on its approved labeling, is not an uncommon practice. "Off-label" use of drugs that are designed for specific subpopulations could be potentially dangerous, and is an issue that must be addressed by regulatory agencies and the medical community.

The potential for pharmacogenomics can be seen today in the use of Genentech's drug Herceptin to treat a genetically distinct form of breast cancer. A fluorescence in situ hybridisation (FISH) test can detect HER2 gene amplification in breast cancer cells, which identifies the approximately 35% of women for whom Herceptin will be effective. In August of 2002, the FDA approved the inclusion of information about the FISH test produced by Abbott on Herceptin's labeling.

Gene Therapy

Another area of medical exploration that is expected to grow as a result of the sequencing of the human genome is gene therapy. Researchers hope to be able to transfer a normal gene into cells to correct a specific disorder. Gene therapy is essentially treating a disease from the inside out. Instead of inhibiting enzymes in certain metabolic pathways and receptors on cells as traditional pharmaceuticals do, gene therapy goes inside and produces the proteins that the cell naturally produces. Some diseases being targeted for treatment by gene therapy are genetic disorders, cancer and HIV/AIDS.

The two primary concerns in gene therapy are introducing genes into specific cells and treating the disorder without aggravating the immune system or inactivating important genes. Most current gene therapies use viral vectors to introduce the gene into specific cells in the body. Retroviral vectors can target genes into human cells and integrate into the host cell. However, the integration can occur randomly and retroviruses only target actively dividing cells. Since the retroviral insertion is random, it could inactivate important genes such as tumor suppressors or activate oncogenes.

Adenoviruses are another possible vector for gene therapy since they target nondividing cells and the foreign gene is highly expressed, producing larger amounts of protein. The adenovirus also does not integrate into the genome, so mutation of the human genome is unlikely. Some of the drawbacks to using adenoviruses in gene therapy are that it can cause an inflammatory immune response and the response is not as targeted since does not integrate into the genome. Minimising the amount of viral DNA in future adenoviral vectors will hopefully decrease the adverse immune response while retaining the therapeutic characteristics of the treatment.

Some nonviral gene therapies are also being explored. In transkaryotic therapy, an individual's skin cells are collected and the gene is inserted into them. The cells are then cultured to increase their number and then they are injected under the patient's skin. Other techniques under consideration are: electroporation, liposome transfer, microinjection, and microinjectile bombardment. Gene therapy is promising technology, but has recently had a number of setbacks in both the United States and Europe.

Future Medicines

One of the goals of the biomedical industry is to discover therapies to treat the underlying cause of disease rather than treat the symptoms of disease, which is what current pharmaceuticals do. While the sequencing of the human genome will contribute to achieving this goal, many other advances in technology will be required.

The Pharmaceutical Research and Manufacturers of America (PhRMA) think that genetic research will change the practice of medicine from treating to preventing disease. The shift in patient care from disease treatment to disease prevention decreases the cost of heath care dramatically. Though in its early stages, genetic research provides a realistic hope for a new and better approach to health care in the 21century.

Ethical Social Issues

The range of ethical issues in biotechnology varies from the treatment of animals, to privacy and discrimination, to the possibility of cloning human beings. Using animal models to test new drugs is an established practice and is part of the preclinical trial stage for drug approval in the U. S. Some people and groups, however, strongly object to animal testing. Currently a number of genetic tests for diseases such as breast and colon cancer exist. These tests can help patients receive earlier monitoring with the hope of detecting the disease at an earlier and more treatable stage. But many patients do not have these tests run because they fear discrimination from their health insurance provider.

Legislation to prevent discrimination based on predictive genetic information is being developed. Many people have ethical concerns about recombinant DNA, human embryonic stem cell, germ-line gene therapy, and cloning technologies. The ethical arguments for and against pursuing biotechnology are numerous, but two very important ones are that its benefits will not be distributed equally and that it promotes the practice of eugenics. Some of the distributive issues are:

— Whether it is just for only those who can afford genetic services to have access to them, especially since much of the initial research that led to these services was publicly funded

— Whether the right to health care includes entitlements to genetic enhancements as well as treatment and prevention of diseases

— Whether international distributive justice requires that the fruits of genetic science (in medicine and agriculture in particular) be shared with those in poorer countries that cannot afford to develop the technology, as opposed to leaving the distribution of these benefits to the global market

— Whether the direction of genetic research and development should be shaped by expected market demand, as opposed to having ethical principles determine priorities (e. g., channeling funds to cure devastating diseases that afflict large numbers of people rather than to the enhancement of normal traits for the rich or the discovery of treatments for less serious conditions, such as male baldness).

Some members of the disability rights community view biotechnology as a tool for excluding individuals with "bad genes" from the respect and consideration that all humans deserve, including life itself through selective abortion. Many people also have concerns about germ-line gene therapy, which would change the genetic make-up of future generations. Other groups, particularly those supporting the right to life movement, object to using human embryonic stem cells in medical technologies because they believe it involves the destruction of a life. They also oppose all human cloning, including therapeutic cloning, since they feel it is creating life to destroy it. Regulatory agencies around the world will have to deal with these as well as many other ethical issues when creating policy. Developing international guidelines would be very difficult since values and ethics vary from person to person and from country to country.

Regulatory agencies

Biotechnology products are classified in a variety of categories including new plant varieties, drugs, biologics, and novel organisms, causing the regulation of the products to fall under the auspices of the different agencies. All biotechnology products are required to meet the same rigorous safety standards that all other agricultural and pharmaceutical products must meet. In the United States, three government agencies currently oversee the regulation of biotechnology: the Food and Drug Administration (FDA), the Department of Agriculture (USDA), and the Environmental Protection Agency (EPA).

In Europe, each country has its own pharmaceutical regulatory agency, but a European Union system for authorisation of medicinal products was established in 1995. This system offers two methods for authorising medicinal products. A company can apply directly to the European Agency for the Evaluation of Medicinal Products, also known as the European Medicines Evaluation Agency or EMEA, which leads to the granting of a European marketing authorisation by the European Commission. All pharmaceutical products derived from biotechnology must be authorised using this procedure. Other medicinal products can be approved under a "mutual recognition" procedure.

The company applies to the member states and the procedure operates by mutual recognition of national marketing authorisations. In Japan, the Ministry of Health and Welfare oversees the regulation of pharmaceuticals. Guidelines for Manufacturing Drugs by Recombinant DNA Technology were established in 1986. The guidelines specify four levels of safety for recombinant products, the establishment of an institutional biosafety committee, the appointment of a biological safety officer, and supervision by a product security pharmacist. Additional notifications have been made to address new biotechnology issues, such as cell culture and gene therapy.

Qualityof Pharmaceuticals

All of these agencies are responsible for setting for laws, regulations, and guidelines to ensure and assess the safety, quality, and efficacy of new medicinal products. Some safety issues include the identification of a safe dose size, any potential targets for toxicity and if it is reversible, and parameters for clinical monitoring. Drug quality is determined by the stability of the drug, the level of impurity, and the analytical validation of the compound. The efficacy of a drug is determined by how well it works to treat the symptoms or disease. All three of these properties are tested in clinical trials.

To determine if a new drug is reasonably safe to test on humans, a company or research institution must perform preclinical testing in lab animals. This, along with the basic discovery work takes an average of six-and-a-half years. When a sponsor of a new drug decides that it would like to begin clinical trials in humans in the United States, they must submit an Investigational New Drug Application to the FDA, which includes the results of the basic research and animal testing and describes the human testing that it proposes to do. The protocol must specify the type of people who will participate in the trial, the drug and dosages used, and the length and goals of the study.

A local institutional review board that is made up of scientists and nonscientists from hospitals and research institutions then reviews the proposed protocols. In the first phase of clinical trials, called Phase I toxicity testing, researchers are looking at drug safety and are trying to determine probable safe dosages. They also are trying to determine the most frequent side effects and learn how the drug is metabolised and excreted. To carry out the Phase I trials, the drug is generally administered to 20 to 80 healthy volunteers for about a year and a half. If the results of the Phase I trials demonstrate an acceptable level of toxicity, the sponsor can begin Phase II trials that determine the efficacy of the drug. Only about two out of five drugs tested in Phase I trials progress to Phase II.

The drug is given to 100 to 500 volunteer patients suffering from the appropriate disease for approximately two years. Of the drugs tested in Phase II trials, only about half emerge as successful, meaning that it effectively treats the disease. During Phase III, which lasts about three-and-a-half years, the drug is given on average to 1,000 to 5,000 volunteer patients. Most of the drugs that are successful in Phase II trials are also successful in Phase III trials. Once the first three phases of clinical trials are complete, the drug sponsor can submit a market application (New Drug or Biologic License Application) to FDA. This includes all animal and human data, data analysis, information on how the drug acts in the body, and manufacturing information.

The drug sponsor must prove that the new drug is safe and effective, and that its benefits outweigh its risks. Based on the application, FDA will decide whether or not to approve the drug. The entire process from drug discovery to approval can take ten to 15 years and costs an estimated $800 million. After FDA approval, the drug sponsor can begin Phase IV trials to explore new uses for the drug, new patient populations, and the long-term effects of the drug.

In addition, the U. S. Pharmacopoeia may establish quality standards for the ingredient and dosage form. These standards are published in the United States Pharmacopeia-National Formulary (USP-NF), and help to assure product identity, strength, quality, and purity. Other countries have similar processes in place for drug testing and approval to ensure the safety, quality, and efficacy of drugs.

Intellectual Property Protection

Some specific policies passed during the 1980s in the U. S. have an important impact on the biotechnology industry. The U. S. Congress passed the Bayh-Dohl act in

1980, which permitted publicly funded researchers and organisations to patent their discoveries. Bayh-Dohl was intended to facilitate the commercialisation of discoveries made in universities and public institutions. The Hatch-Waxman Act was designed to allow pharmaceutical companies to run abbreviated studies in order to increase the availability of therapeutically equivalent second-entry (generic) drug products after patent and/or exclusivity protection expired.

Because of this law, the generic drug industry now holds almost 50% of the U. S. prescription drug market, as opposed to the less than 19% it held before the law was passed. Unfortunately, the act did not distinguish between drug patents and research tool patents, and a number of companies have successfully defended themselves from patent infringement involving research tools using the Hatch-Waxman Act. The impact of this legislation could be severe, since research tools are a vital part of a biotechnology firm's intellectual property.

More recent intellectual property developments could have great consequences for the biotechnology and pharmaceutical industries. In December 2000, the U. S. Patent and Trademark Office established its guidelines for the patentability of human genes, requiring applicants requesting patents for gene sequences to demonstrate the specific, substantial, and credible utility of the sequence. On the international level, the World Trade Organisation (WTO) has also set forth some guidelines for pharmaceutical, and therefore biotechnology, patents in its Agreement on Trade-Related Aspects of Intellectual Property Rights (TRIPS).

WTO members are required to provide patent protection for all inventions, both products and processes, for at least 20 years from the date the patent application was filed. However, there are some exceptions.

Governments can refuse to grant patents if they cover inventions that can be used to protect human, animal or plant life or health; are diagnostic, therapeutic, and surgical methods for treating humans or animals; or are certain plant and animal inventions. The TRIPS Agreement also provides "other use without authorisation of the right holder", also known as compulsory licensing, which allows a government to permit someone else to produce a patented product or process without the consent of the patent owner. This can only be done when a person or company applying for a license has been denied a voluntary license from the right holder in the cases of "national emergency," "other circumstances of extreme urgency," "public non-commercial use," or anti-competitive practices.

Finally, "The WTO council responsible for intellectual property, on 27 June 2002, approved a decision extending the transition period during which least-developed countries (LDCs) do not have to provide patent protection for pharmaceuticals until 2016. The Council for Trade-Related Aspects of Intellectual Property Rights also approved a waiver that would exempt least-developed countries from having to provide exclusive marketing rights for any new drugs in the period when they do not provide patent protection."

However, LDCs are still required to allow inventors to submit patent applications during this time period, and if a new drug is approved for sale, the patent applicant must have exclusive marketing rights for five years. The Council made this decision in response to the epidemics afflicting many LDCs including, HIV/AIDS, tuberculosis, and malaria. The outcomes of these patent policies will be seen over the next decade or two, but they are designed to provide protection of intellectual property rights while also putting global public health concerns, such as the HIV/AIDS epidemic, as a priority.

Generic Drugs

The patents for many of the first recombinant DNA and biotechnology pharmaceutical products have recently expired or will do so in the near future. Analysts have estimated that recombinant products worth more than $10 billion in annual sales will lose patent protection by 2006. Generic pharmaceutical companies are looking to have their biologics approved in a manner similar to the way that more traditional generic pharmaceuticals are approved. Currently, generic biologics must go through the expensive and meticulous clinical trial process, unlike generic small-molecule drugs that can be approved on the basis of bioequivalence data only.

The generic drug companies claim that the clinical trials for generic biologics give biotechnology companies unfair monopolies on their products even after the patents expire. The Biotechnology Industry Organisation is working to keep the clinical trials in place for generic biologics due to safety concerns. "Unlike traditional drugs, biologics are more than just pure chemical preparations: small changes to the manufacturing process can lead to clinical differences in the way products work. That makes it impossible to tell whether the same protein made by two different manufacturers will work in the same way".

For example, Johnson & Johnson and Amgen both produce versions of erythropoietin that are marketed in Europe. The Johnson & Johnson drug, Eprex has been linked to an estimated 250 cases of pure red-cell aplasia, a rare cause of anemia, while the Amgen drug, Epogen has been linked to a similar problem only once. The two molecules made by two different companies appear to have different effects on the body despite both being a form of same protein. Scientists have found that aggregates of small molecules that have resulted from the substitution of sorbitol 80 in place of Human Serum Albumin were responsible for the severe immunogenic reaction triggered the severe side effects.

Sorbitol 80 was used as the stabiliser in the product rather than Human Serum Albumin to avoid potential contamination by HIV and Creutsfeldt-Jacob disease-causing prions. Eprex has provided an example of how a change in formulation of a therapeutic protein can have a significant impact. All therapeutic proteins have the potential to induce immunogenic reactions, which could be triggered not only by a change in formulation, but also by variations in amino acid sequence, glycosylation, or impurities occurring during the manufacturing or administration of the drug.

The generic drug industry claims that the challenges of creating generic recombinant drugs and other biologics can be overcome, and the FDA has drafted regulation for a few older recombinant products, such as insulin, to be approved without going through the full clinical trial process. Further regulations allowing for generic biologics will likely be passed because of the high price of biotechnology products. Production of generic biologics not required to go through the extensive clinical trial process would allow the price of biotech drugs to decrease. However, the issue of safety is always a priority in regulation.

Harmonisation

As the pharmaceutical industry continues to become more global, the need to harmonise regulation has become apparent. While the various national regulatory systems have had similar goals to evaluate and ensure the safety, quality and efficacy of medicinal products, the detailed technical requirements have diverged over time to such an extent that companies are required to duplicate many expensive and time-consuming procedures in order to market new products internationally. The cost savings of a harmonised drug review process are expected to be pretty significant. Harmonisation would also minimise the use of animal testing without compromising the regulatory obligations of safety and effectiveness.

The first harmonisation efforts took place in the European Community during the 1980s to create a single European market for medicinal products. This initiative was shortly followed in 1990 by the International Conference on Harmonisation of Technical Requirements for Registration of Pharmaceuticals for Human Use (ICH) which was established by representatives of the regulatory authorities and industry associations of Europe, Japan, and the United States. They are: the European Union (EU), the European Federation of Pharmaceutical Industries' Associatiosns (EFPIA), the Ministry of Health, Labor and Welfare, Japan (MHLW), the Japan Pharmaceutical Manufacturers Association (JPMA), the U. S. Food and Drug Administration (FDA), and the Pharmaceutical Research and Manufacturers of America (PhRMA).

Since the majority of new drugs are produced in these three regions, ICH has initially confined its efforts to them. Other countries and organisations are, however, observers of the ICH process, including the World Health Organisation (WHO), the European Free Trade Area (EFTA), and Canada. The ICH Secretariat is run by the International Federation of Pharmaceutical Manufacturers Association (IFPMA). The ICH Steering Committee has decided that the topics selected for harmonisation would be safety, quality, and efficacy since they are the three criteria used for approving new pharmaceuticals. Expert Working Groups (EWGs) were set up to discuss the scientific and technical aspects of each harmonisation topic. Three principles of ICH work were also established. They are:

— Development of scientific consensus through discussions between regulatory and industry experts;
— Wide consultation of the draft consensus documents through normal regulatory channels before a harmonised text is adopted; and
— Commitment by regulatory parties to implement the ICH harmonised texts

The goal of the first phase of ICH activities, which was completed in July 1997, was to remove redundancy in the review and registration process so that one set of data could be generated to demonstrate the safety, quality, and efficacy of a new drug. As a result, the ICH Guideline E3, Content and Format of Clinical Study Reports, and The Common Technical Document (CTD) were created. Guideline E3 outlines the procedures for a single format for reporting the core clinical studies that make up the clinical section of a registration dossier. The CTD provides a harmonised format for new product applications and authorisations.

In the second phase of ICH activities, the Steering Committee has established a mechanism to harmonise (a) new technical requirements that emerge as scientific developments occur and (b) a process for monitoring and updating current ICH guidelines. Through early collaboration and exchange of information on new issues, they hope to prevent future conflicts among the three regions. The Steering Committee also expects the second phase to cover a wider range of issues that may have implications for parties outside the founding members of ICH. In order to disseminate the ICH guidelines and encourage their adoption by countries outside of the three original regions, ICH is working with WHO, which will hopefully result in faster access to new pharmaceuticals throughout the world.

ICH has addressed the evolving issue of biotechnology in the development of new drugs in some of the guidelines that have been finalised or are in the process of finalisation. Workshops on important biomedical technologies with potential for harmonisation, such as gene therapy, have also been held. Most of the guidelines that specifically address biotechnology products fall under the topics of safety and quality.

The finalised safety guideline is S6: Preclinical Safety Evaluation of Biotechnology Derived Pharmaceuticals, which is intended to recommend a basic framework for preclinical safety evaluation for active substances that are derived from cell cultures or produced using recombinant DNA technology including production by transgenic plants and animals. This includes determining safe dose size, toxicity, and safety parameters.

Biotechnology has already provided major advancements in medical technology and the future promises to hold even more. These advances will pose many challenges to the scientific, medical, and regulatory communities-as well as the general public-who must continue to balance the different priorities that each brings to the debate about biomedical technology. The first priority for all parties, however, should be safety. The United States, the European Union, and Japan have rigorous approval processes for new pharmaceuticals that help to ensure the safety of these products, and are working to harmonise these processes through ICH because of the global nature of the market.

These efforts should continue, but may need to be more inclusive since some countries have already voluntarily adopted the ICH guidelines and the biotechnology industry has been established in many countries where a research-oriented pharmaceutical industry did not previously have a significant presence.

The ICH guidelines to ensure the quality of biotechnology products are an important step forward in guaranteeing the safety of patients that use approved biotechnology drugs, and ICH should continue to create new quality guidelines as the advancement of the technology requires.

Bibliography

Adam, G. and H. Duncan. *Effect of Diesel Fuel on Growth of Selected Plant Species*. Environ. Geochem. Health 21(4): 353-357, 1999.

Adler, P.A., R. Arora, A. El Ghaouth, D.M. Glenn, and J.M. Solar. *Bioremediation of phenolic compounds from water with plant root surface peroxidases*.J. Environ. Qual. 23:1113-1117, 1994.

Ahlf, W., W. Calmano, J. Erhard, and U. Fostner. "Comparison of five bioassay techniques for assessing sediment-bound contaminants". *Hydrobiologia*. 188-189:285-290, 1989.

Anderson, J.W., and A.R. Scarf. "Selenium and plant metabolism". pp. 241-275. In D.A. Robb and W.S. Pierpoint (eds.), Metals and micronutrients: Uptake and utilization by plants. Academic Press, New York, NY, 1983.

Anderson, P. "These plants sop up metal pollutants". *Business Week*. March 9. p. 103, 1987.

Anderson, T.A. "Comparative plant uptake and microbial degradation of trichloroethylene in the rhizospheres of five plant species: Implications for bioremediation of contaminated surface soils", 1991

Aranda, J.M., G.A. O'Connor, and G.A. Eiceman. Effects of sewage sludge on diethylhexyl phthalate uptake by plant., J. Environ. Qual. 18:45-50, 1989.

Banks, M.K., A.P. Schwab, G.R. Fleming, and B.A. Hetrick. Effects of plants and soil microflora on leaching of zinc from mine tailings. Chemosphere. 29:1691-1699, 1994.

_______________, J.S. Smith, K. Miller, K. Rathbone, and P. Kulakow. Bioavailability of Petroleum Contaminants in Vegetated Soil. (1998), 14th Annual Conference on Contaminated Soils. October 1998. University of Massachusetts at Amherst, Amherst, MA..

Banuelos, G.S., and D.W. Meek. *Selenium accumulation in selected vegetables.*, J. Plant Nutr. 12(10):1255-1272, 1989.

_______________, R.R. Mead, and G.J. Hoffman. "Accumulation of selenium in wild mustard irrigated with agricultural effluent". *Agric. Ecosystems Environ.* 43:119-126, 1993.

Banuelos, G.S., R.R. Mead, and S. Akohoue. *Adding selenium-enriched plant tissue to soil causes the accumulation of selenium in alfalfa.* J. Plant Nutr. 14(7):701-713, 1991.

Sandermann, H. *Plant Metabolism of Xenobiotics.* TIBS 17: 82 - 84, 1992.

Sandermann, H. "Higher Plant Metabolism of Xenobiotics: The 'Green Liver' Concept"., *Pharmacogenetics* 4: 225-241, 1994.

Soudel, P., R. Podlipna, H. Lipavska and T. Vanek. Bioaccumulation of Heavy Metals by Hair-Root Cultures of Armoracia Rusticana L. (I.). *Pharmaceutical Pharmacol.* Letters 8(2): 57-60, 1998.

Index